Arduino-Based Embedded Systems

Arduino-Based Embedded Systems

Interfacing, Simulation, and LabVIEW GUI

Rajesh Singh

Anita Gehlot

Bhupendra Singh

Sushabhan Choudhury

CRC Press
Taylor & Francis Group
Boca Raton London New York

CRC Press is an imprint of the
Taylor & Francis Group, an **informa** business

CRC Press
Taylor & Francis Group
6000 Broken Sound Parkway NW, Suite 300
Boca Raton, FL 33487-2742

International Standard Book Number-13: 978-1-1380-6078-4 (Hardback)
International Standard Book Number-13: 978-1-315-16288-1 (ebook)

Library of Congress Cataloging-in-Publication Data

Names: Singh, Rajesh (Electrical engineer), author. | Gehlot, Anita, author.
| Singh, Bhupendra, author. | Choudhury, Sushabhan, author.
Title: Arduino-based embedded systems : interfacing, simulation, and LabVIEW
GUI / Rajesh Singh, Anita Gehlot, Bhupendra Singh and Sushabhan Choudhury.
Description: Boca Raton : Taylor & Francis, CRC Press, 2018.
Identifiers: LCCN 2017029926 | ISBN 9781138060784 (hardback : alk. paper) |
ISBN 9781315162881 (ebook)
Subjects: LCSH: Embedded computer systems--Programming. | Arduino
(Programmable controller)--Programming. | LabVIEW.
Classification: LCC TK7895.E42 S548 2018 | DDC 006.2/2--dc23
LC record available at https://lccn.loc.gov/2017029926

Visit the Taylor & Francis Web site at
http://www.taylorandfrancis.com

and the CRC Press Web site at
http://www.crcpress.com

Contents

Section III   Arduino and Wireless Communication

Section IV   Projects

Preface

The primary objective of writing this book is to provide a platform for beginners to get started with the Arduino-based *embedded system* and who need a basic knowledge of programming and interfacing of the devices.

The aim of this book is to explain the basic steps to get started with the Arduino and to develop an understanding of the interfacing, programming, and simulation of the designed systems.

This book comprises 25 chapters and is divided into 4 sections. *Section I* of this book is about the introduction to the basic software that is required to get started with the Arduino. *Section II* is about interfacing of display devices and basic input/output devices such as sensors and motors. *Section III* is about the interfacing of basic communication modules such as RF modem and global system for mobile (GSM). *Section IV* includes examples of Arduino-based projects. This book is intended to serve the students of B.Tech/B.E, M.Tech/M.E, PhD scholars, and who need the basic knowledge to develop a real-time system using the Arduino.

We acknowledge the support from Sunrom technologies, Robosoft systems, and Robokits India for using their product images and data to demonstrate and explain the working of the systems. We thank Taylor & Francis/CRC Press for encouraging our idea about this book and the support to efficiently manage the project.

We are grateful to the honorable chancellor Dr. S.J Chopra, Utpal Ghosh (President & CEO, UPES), Dr. Srihari (Vice-chancellor, University of Petroleum and Energy Studies (UPES)), Dr. Kamal Bansal (Dean, CoES, UPES), Dr. Suresh Kumar (Director, UPES), and Dr. Piyush Kuchhal (Associate Dean, UPES) for their support and constant encouragement. In addition, we are thankful to our families, friends, relatives, colleagues, and students for their moral support and blessings.

Although the circuits and programs mentioned in the text are tested on real hardware but in case of any mistake we extend our sincere apologies. Any suggestions to improve in the contents of the book are always welcome and will be appreciated and acknowledged.

Rajesh Singh
Anita Gehlot
Bhupendra Singh
Sushabhan Choudhury

About the Authors

Dr. Rajesh Singh is currently associated with the University of Petroleum and Energy Studies, Dehradun, India, as an associate professor and with additional responsibility as Head, Institute of Robotics Technology (R&D). He has been awarded a gold medal in MTech and honors in his BTech. His area of expertise includes embedded systems, robotics, and wireless sensor networks. He has organized and conducted a number of workshops, summer internships, and expert lectures for students as well as faculty. He has 12 patents in his account. He has published approximately 100 research papers in refereed journals/conferences.

Under his mentorship, students have participated in national/international competitions, including Texas Instruments Innovation Challenge in Delhi and Laureate award of excellence in robotics engineering in Spain. Twice in the last 4 years he has been awarded with a certificate of appreciation from the University of Petroleum and Energy Studies for his exemplary work. He received a certificate of appreciation for mentoring the projects submitted to the Texas Instruments India Innovation Challenge Design Contest 2015 from Texas Instruments. He has been honored with a young investigator award at the International Conference on Science and Information in 2012. He has published a book titled *Embedded System based on Atmega Microcontroller* with the NAROSA publishing house. He is an editor to a special issue published by the Advances in Intelligent Systems and Computing (AISC) book series, Springer titled *Intelligent Communication, Control and Devices 2016*.

Anita Gehlot has more than 10 years of teaching experience with an area of expertise in embedded systems and wireless sensor networks. She has 10 patents in her account. She has published more than 50 research papers in both refereed journals and conferences. She has organized a number of workshops, summer internships, and expert lectures for students. She has been awarded with a certificate of appreciation from the University of Petroleum and Energy Studies, Dehradun, India, for her exemplary work. She has coauthored a book titled *Embedded System based on Atmega Microcontroller* with the NAROSA publication house.

Bhupendra Singh is the managing director of Schematics Microelectronics and provides product design and R&D support to industries and universities. He has completed BCA, PGDCA, MSc (CS), MTech, and has more than 11 years of experience in the field of computer networking and embedded systems.

Dr. S. Choudhury is the head of the Department of Electronics, Instrumentation, and Control at the University of Petroleum and Energy Studies, Dehradun, India. He has 26 years of teaching experience and he earned his PhD from the University of Petroleum and Energy Studies, MTech (Gold Medalist) from Tezpur Central University, Tezpur, India, and earned his BE degree from National Institute of Technology Silchar, India. He has published more than 70 papers in various national/international conferences/journals and has filed 10 patents. His area of interest is Zigbee-based wireless networks. Dr. Choudhury has been selected as the outstanding scientist of the twenty-first century by the Cambridge Biographical Centre, Cambridge, UK. He has also been selected in the who's who of the world in science by Marquis Who's Who, New Providence, New Jersey. He has coauthored a book titled *Embedded System based on Atmega Microcontroller* with the NAROSA publishing house. He is an editor to a special issue published by the AISC book series, Springer, titled *Intelligent Communication, Control and Devices 2016*.

Section I

Introduction

1

Introduction to Arduino

Arduino is a user-friendly open-source platform. Arduino has an onboard microcontroller and an integrated development environment (IDE) is used to program it. Arduino board can be programmed directly from the PC using FTDI which is easy compared to other similar platforms.

The advantages are as follows:

Low cost: Arduino boards are of relatively low cost as compared to other microcontroller platforms.

Cross-platform: The Arduino software (IDE) is compatible with the Windows, Macintosh OSX, and Linux operating systems.

User friendly: The Arduino software (IDE) is user friendly and easy to use for beginners and very flexible for skilled programmers.

Open source: The Arduino is an open-source software and can be programmed with C, C++, or AVR-C languages. So, a variety of modules can be designed by the users.

Arduino platform comprises a microcontroller. It can be connected to a PC through a USB cable. It is freely accessible and can be easily downloaded. It can also be modified by a programmer. Different versions of Arduino boards are available in the market depending on the user requirement.

1.1 Arduino Uno

The Arduino/Genuino Uno has an onboard ATmega328 microcontroller. It has onboard six analog input ports (A0–A5). Each pin can operate at 0–5 V. It has 14 digital input/output (I/O) pins out of which 6 are PWM output, 6 analog inputs, 2 KB SRAM, 1 KB EEPROM, and operates at 16 MHz of frequency (Figure 1.1 and Table 1.1).

FIGURE 1.1
Arduino Uno board.

TABLE 1.1

Pin Description of Arduino Uno

Pin	Description
Vin	The external voltage to the Arduino board
+5 V	+5 V regulated output
3.3 V	On board 3.3 V supply
GND	Ground
IOREF	Provides the voltage reference and select appropriate power source
Serial	Transmits and receives serial data, Pins: 0(Rx) 1(Tx)
External interrupts	Trigger an interrupt on low value, Pins: 2 and 3
PWM	Provides 8 bit PWM output, Pins: 3,5,6,9,10,11
SPI	Supports SPI communication, Pins: 10 (SS), 11 (MOSI), 12 (MISO), and 13 (SCK)
LED	LED driven by Pin 13
TWI	Supports TWI communication, Pins: A4 (SDA), A5 (SCL)
AREF	Reference voltage for the analog inputs
Reset	It is used to reset the onboard microcontroller

1.2 Arduino Mega

The Arduino Uno has onboard ATmega2560 microcontroller. It has onboard 16 analog inputs, 54 digital I/O, USB connection, 4 UART, power jack, and a reset button. It operates at 16 MHz frequency. The board can be operated with 5–12 V of external power; if supplied more than this, it can damage the board. It has onboard 256 KB flash memory, 8 KB SRAM, and 4 KB EEPROM (Figure 1.2 and Table 1.2).

FIGURE 1.2
Arduino Mega board.

TABLE 1.2

Pin Description of Arduino Mega

Pin	Description
Vin	The external voltage to the Arduino board
+5 V	+5 V regulated output
3.3 V	Onboard 3.3 V supply
GND	Ground
IOREF	Provides the voltage reference and select an appropriate power source
Serial0	Transmits and receives serial data, Pins: 0(Rx) 1(Tx)
Serial1	Transmits and receives serial data, Pins: 19(Rx) 18(Tx)
Serial2	Transmits and receives serial data, Pins: 17(Rx) 16(Tx)
External interrupts	Trigger an interrupt on low value, Pins: 2 (interrupt0), 3 (interrupt1), 18 (interrupt5), 19 (interrupt4), and 20 (interrupt2)
PWM	Provides 8 bit PWM output, Pins: 2–13 and 44–46
SPI	Supports SPI communication, Pins: 53 (SS), 51 (MOSI), 50 (MISO), and 52 (SCK)
LED	LED driven by Pin 13
TWI	Supports TWI communication, Pins: 20 (SDA), 21 (SCL)
AREF	Reference voltage for the analog inputs
Reset	It is used to reset the onboard microcontroller

1.3 Arduino Nano

The Arduino/Genuino Nano has onboard ATmega328 microcontroller. It has onboard 8 analog and 14 digital I/O ports and 6 PWM of 8 bit. Each pin can operate at 0–5 V. It has onboard 32 KB flash memory, 2 KB SRAM, 1 KB EEPROM, and operates at 16 MHz of frequency (Figure 1.3 and Table 1.3).

FIGURE 1.3
Arduino Nano board.

TABLE 1.3

Pin Description of Arduino Nano

Pin	Description
Vin	The external voltage to the Arduino board
+5 V	+5 V regulated output
3.3 V	Onboard 3.3 V supply
GND	Ground
IOREF	Provides the voltage reference and select an appropriate power source
Serial	Transmits and receives serial data, Pins: 0(Rx) 1(Tx)
External interrupts	Trigger an interrupt on low value, Pins: 2 & 3
PWM	Provides 8 bit PWM output, Pins: 3,5,6,9,10,11
SPI	Supports SPI communication, Pins: 10 (SS), 11 (MOSI), 12 (MISO), and 13 (SCK)
LED	LED driven by Pin 13
I2C	Supports TWI communication, Pins: A4 (SDA), A5 (SCL)
AREF	Reference voltage for the analog inputs
Reset	It is used to reset the onboard microcontroller

2

Steps to Write a Program with Arduino Integrated Development Environment

This chapter describes the steps to write and compile a program with Arduino integrated development environment (IDE). The Arduino IDE is an open-source software which makes it user friendly for writing the code and then upload directly on Arduino board.

2.1 Steps to Install Arduino Integrated Development Environment

Step 1: Install Arduino IDE and open the Window

To begin, install the Arduino Programmer, IDE. Figure 2.1 shows the opened window Arduino IDE.

Step 2: Choose suitable version of Arduino

Arduino has many versions such as Uno, Mega, and Nano. The most common is the Arduino Uno. Before starting the program find out the suitable version of Arduino board for the project. Set the board type and the USB serial port of board in the Arduino IDE. Figure 2.2 shows the steps to select the type of Arduino. Click on "Tool," and then click on "board." Figure 2.2 shows the selection of "Arduino Uno."

Step 3: Write and compile the program

Write program as per requirement of the project. Then "RUN" the program. Figure 2.3 shows the compilation of the program.

FIGURE 2.1
Window Arduino IDE.

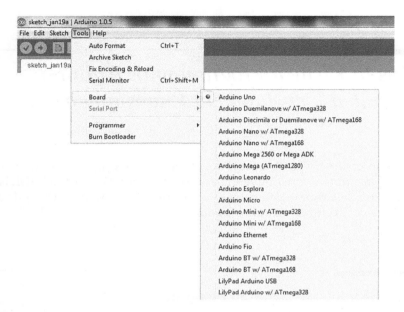

FIGURE 2.2
Window to select type of Arduino.

FIGURE 2.3
Compile the program.

Step 4: Connect Arduino with the PC

Connect Arduino to the USB port of the PC with USB cable. Every Arduino has a different serial-port address (e.g., COM2 and COM4), so it is required to recognize the port for the different Arduino and select it in the IDE.

To check the port where the Arduino is connected, make right click on the "PC," then go to manager; a window will open. Then double click on the "Device Manager." A window as shown in Figure 2.4 will open. Click on the ports (COM&LPT) and the port at which the device is connected can be found. Figure shows "COM6" is port for the device.

Now click on the "Tool" heading of the Arduino IDE window. Go to port and select the same port number, which was found at the device manager (select COM1 or COM2 etc.). Figure 2.5 shows the "COM6" as serial port of board.

Step 5: Upload program in Arduino

Uploaded the new program to Arduino. Figure 2.6 shows how to upload the program.

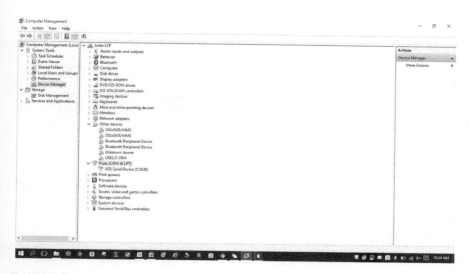

FIGURE 2.4
Window to check port of Arduino.

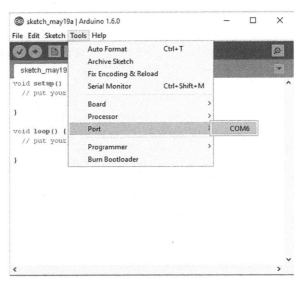

FIGURE 2.5
The serial port of board.

FIGURE 2.6
Window to upload the program in Arduino.

2.2 Basic Commands for Arduino

1. *pinMode(x, OUTPUT)*: // assigned pin number x as output pin in which x is the number of digital pin
2. *digitalWrite(x, HIGH)*: // turn ON the pin number x as HIGH or ON in which x is the number of digital pin
3. *pinMode(x, INPUT)*: // assigned pin number x as input pin in which x is the number of digital pin
4. *digitalRead(digital Pin)*: // read the digital pin as 13 or 12 or 11 etc.
5. *analogRead(analog pin)*: // read the analog pin as A0 or A1 or A2 etc.

LCD Commands

1. *lcd.begin(16, 2)*: // initialize LCD 16 * 2 or 20 * 4
2. *lcd.print("UPES")*: // print a string "UPES" on LCD
3. *lcd.setCursor(x, y)*: // set the cursor of LCD at the desired location in which x is the number of COLUMN and y
4. *lcd.print(UPES)*: // print a UPES as an integer on the LCD
5. *lcd.Clear()*: // clear the contents of the LCD

Serial Communication Commands

1. *Serial.begin(baudrate)*: // initialize serial communication to set baud rate to 600/1200/2400/4800/9600
2. *Serial.print("UPES")*: // serial print fixed string with a defined baud rate on the Tx line
3. *Serial.println("UPES")*: // serial print fixed string with a defined baud rate and enter command on the Tx line
4. *Serial.print(UPES)*: // serial print int string with a defined baud rate on the Tx line
5. *Serial.print(UPES)*: // serial print int string with a defined baud rate and then enter command on the Tx line
6. *Serial.Write(BYTE)*: // serial transfer the 1 byte on the Tx line
7. *Serial.read()*: // read 1 byte serial from the Rx line

3

Steps to Design a Proteus Simulation Model

This chapter describes the steps to design a Proteus simulation model for a system and simulate it to check the feasibility of the designed system.

Step 1: Install ISIS software

Install the ISIS software and open the window.

Step 2: Create a New Design

Open the ISIS software. Select a new design from the File menu. Figure 3.1 shows the ISIS window and create a New Design.

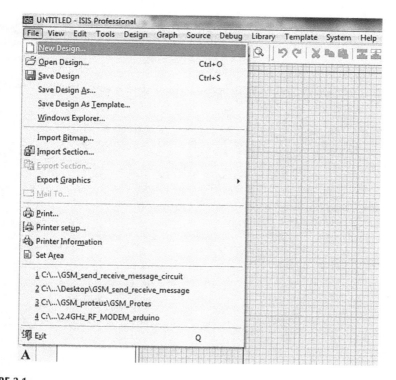

FIGURE 3.1
"New Design" window.

Clicking on "New Design," a dialog box appears to save the current design file. Click on "Yes." Then a pop-up message appears to select the template. Select "Default" template and click on "Yes." Figure 3.2 shows the window for selecting a "default" template.

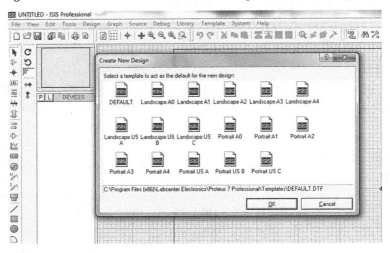

FIGURE 3.2
Select "Default" template.

Step 3: Open window for new model

An untitled design sheet will open; save it under any title. Figure 3.3 shows the new window.

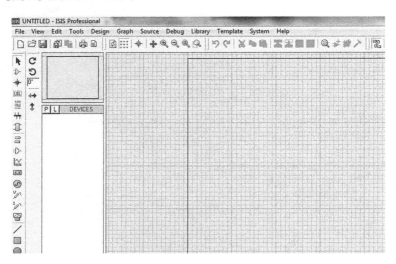

FIGURE 3.3
Window for New Design.

Step 4: Choose Components from the list

To choose the components, click on the component mode. Click on "P" Pick from the libraries as shown in Figure 3.4. It shows the categories of components that are accessible and also a search option to enter the part name, as shown in Figure 3.5. The selected components will appear in the devices list as shown in Figure 3.6. Select the component and place it in the design sheet by a left click.

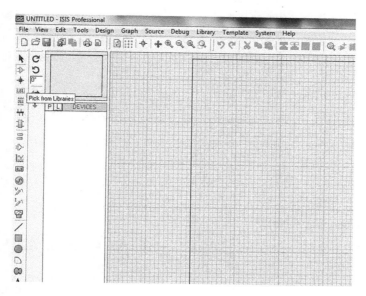

FIGURE 3.4
Window for "Pick from libraries."

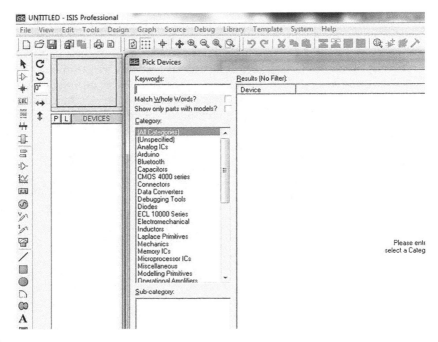

FIGURE 3.5
Select components from the list.

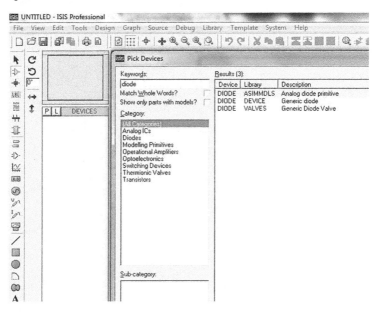

FIGURE 3.6
Window showing selected components.

Step 5: Connect the components

Connect all the components as per the circuit diagram of the system, by clicking on the terminal, which has to be connected with the other component. A wire will appear that will connect them. Figure 3.7 shows a circuit to glow LED by pressing a switch. After connecting the circuit, click on the Play button to run the simulation. Figure 3.8 shows the output of the circuit by clicking on the "Play" button at left most bottom of the window. Click on the switch in the circuit to check the working of the model.

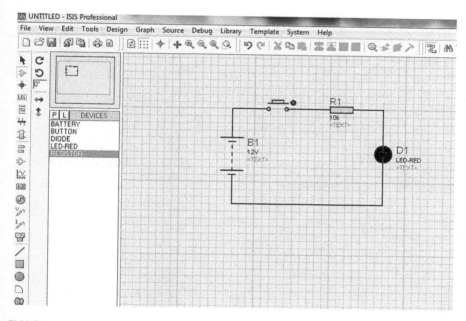

FIGURE 3.7
Proteus simulation model.

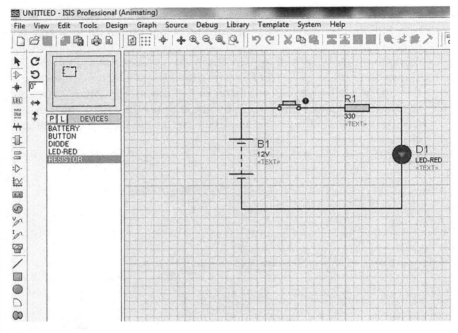

FIGURE 3.8
Proteus simulation working model.

4

Introduction to LabVIEW GUI

This chapter describes the design steps for LabVIEW graphical user interface (GUI) with a brief introduction to the basic blocks that are used to design the GUI. LabVIEW GUI has two components: (1) the front panel and (2) the block diagram. Control can be customized for different work environment. Front panel is a window in which the final GUI will appear and at its back end the "block diagram" runs, which is basically a graphical programming language. Block diagram is designed with some predefined blocks and GUI is developed by connecting the blocks.

4.1 Steps to Design LabVIEW GUI

Step1: Install the LabVIEW software

The first step is to install the LabVIEW software. It is a licensed software and the user need to purchase it from the National Instruments (NI). Install the NI-Virtual Instrument Software Architecture (VISA) driver. NI-VISA driver is required for serial communication, without it data cannot communicate with the serial port of the computer.

Step 2: Create "New Project"

Open the LabVIEW window and click on the "Create Project" button, as shown in Figure 4.1.

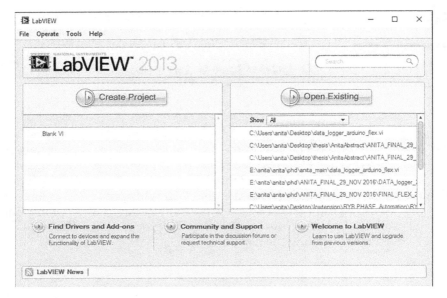

FIGURE 4.1
LabVIEW window to "Create project."

Select Blank VI from the list of items and click on the Finish button
(Figure 4.2). A blank front panel window and block diagram window will appear (Figure 4.3).

Step 3: Running and Debugging VIs

To run a VI, connect all the sub VIs, functions, and structures as per the
data types for the terminals.

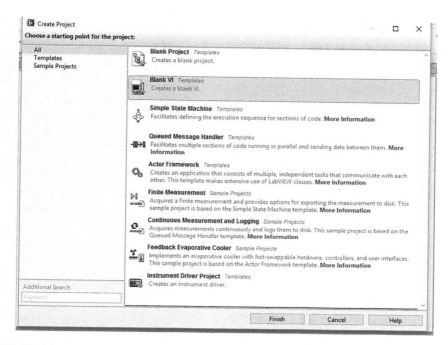

FIGURE 4.2
Select "Blank VI" window.

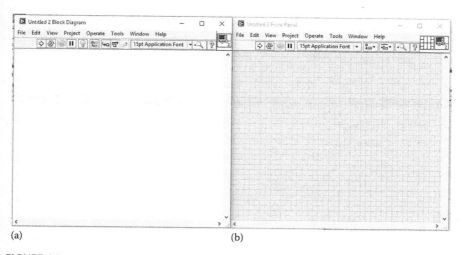

FIGURE 4.3
(a) Blank "block diagram" and (b) "front panel" window.

4.2 Building the Front Panel

The front panel is the user interface of a VI. Generally, the front panel is designed first and then the block diagram is designed to perform tasks on the inputs and outputs that are created on the front panel. To design front panel make as right click anywhere on the window and select the controls and indicators. The selection of controls and indicators can also be done by clicking on *Select View*» select the "Controls Palette" and then select controls and indicators and place them on the front panel. Controls are knobs, push buttons, dials, and other input mechanisms. Indicators are graphs, LEDs, and other output displays. Figure 4.4 shows the window for "front panel" with the control panel and Figure 4.5 shows an example for the front panel.

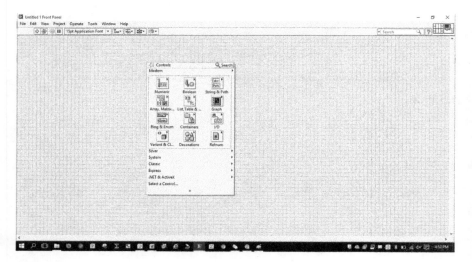

FIGURE 4.4
Window to design "front panel."

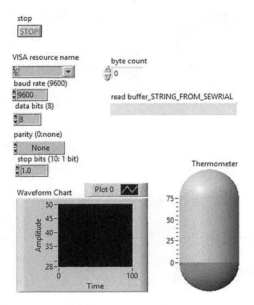

FIGURE 4.5
Example of front panel.

4.3 Building the Block Diagram

Block diagram is basically a code that uses graphical representations of functions to control the front panel objects. The block diagram contains the graphical source code, which is also known as "*G code.*" The components can be selected by a right click on the front panel or go to View and then select "Function Palette," as shown in Figure 4.6.

Figure 4.7 shows an example of block diagram.

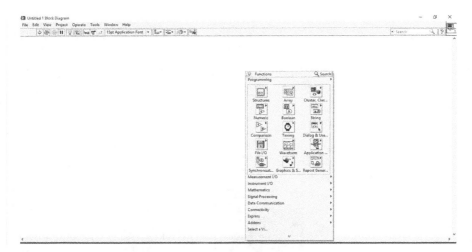

FIGURE 4.6
Select components for "front panel."

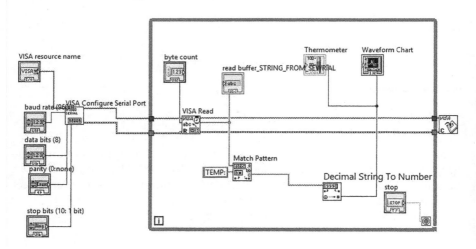

FIGURE 4.7
Example of "block diagram."

4.4 Virtual Instrument Software Architecture

The VISA is a standard, which is used to configure the programming and troubleshooting of instrumentation systems. VISA provides the programming interface between the hardware and LabVIEW. NI-VISA is an implementation of the National Instruments I/O standard. NI-VISA includes software libraries.

4.4.1 Components Used to Design LabVIEW GUI

VISA configure serial port: This port initializes the serial port specified by the VISA resource name with the required settings. Figure 4.8 shows the visa serial port.

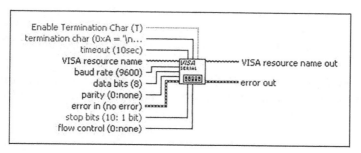

FIGURE 4.8
VISA serial port.

VISA resource name (COM number): Right click on it and select *create*, then select *control* to choose an appropriate COM port.

Baud rate: Choose 9600

Data bits: Choose 8 bits

Parity: Choose none

Stop bit: Choose 1

Flow control: Choose none

VISA resource name out: Connect this pin to *VISA resource name* of the *VISA serial read* block.

Error out: Connect this pin to error pin of the *VISA serial read*.

VISA serial read: It reads the identified number of bytes from the device or an interface identified by the *VISA resource name* and sends the data in the *read buffer*. It reads the data available at the serial port from the device linked (Figure 4.9).

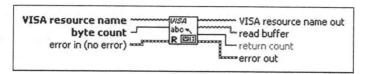

FIGURE 4.9
VISA serial read.

Byte count: Right click on it and select *create* to the *indicator* to count the byte at the serial port.

Read buffer: Right click on it and select *create* to the *indicator* to check the string value at the serial port.

Visa resource name out: Connect this pin to *VISA resource name* of *VISA close* block.

Error out: Connect this pin to the error pin of *VISA close*.

Error in: Connect this pin to error out the pin of *VISA configure serial port*.

Match pattern: It searches for an expression in the string beginning at offset, and when it gets the string it matches the string with the predefined data (Figure 4.10).

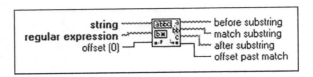

FIGURE 4.10
Match pattern.

String: Connect this pin to read buffer pin of the *VISA read*.

Regular expression: Right click on it and select *create* to the *constant*.

After substring: Connect this pin to the *string* of *Decimal String to Number* block and *create* to *constant*.

Decimal string to number: It converts the numeric characters in the *string*, starting at *offset*, to a decimal integer and returns it in *number* (Figure 4.11).

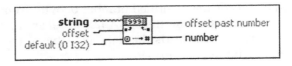

FIGURE 4.11
Decimal string to number converter.

Number: Connect this pin to the input of a *waveform chart*.

VISA close: Closes a device session or event object specified by the VISA resource name (Figure 4.12).

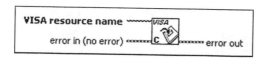

FIGURE 4.12
VISA close.

VISA serial write: Writes the data from write buffer to the device or interface that is identified by the VISA resource name (Figure 4.13).

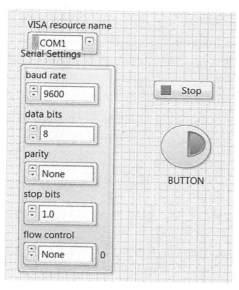

FIGURE 4.13
Defining VISA resource name.

VISA configure serial port: It sets the serial port identified by the VISA resource name to the specified settings (Figure 4.14).

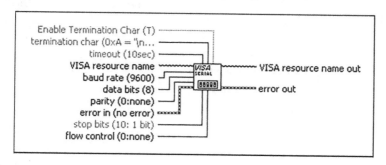

FIGURE 4.14
VISA configure serial port.

VISA resource name (COM number): Right click on it, select *create*, and then select the *control* to choose an appropriate COM port.

Baud rate: Choose 9600

Data bits: Choose 8 bits

Parity: Choose none

Stop bit: Choose 1

Flow control: Choose none

VISA resource name out: Connect this pin to the *VISA resource name* of *VISA serial read* block.

Error out: Connect this pin to error pin of the *VISA serial read*.

VISA serial write: It writes the data from the write buffer to the device or interface as stated by the VISA resource name (Figure 4.15).

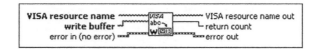

FIGURE 4.15
VISA serial write.

VISA resource name: Connect this pin to the *VISA resource name out* pin of *VISA configure serial port*.

Write buffer: It comprises the data to be written to the device. Right click on it and select *create* to the *constant* to send string at the serial port.

VISA resource name out: Connect this pin to the *Visa resource name* of the *VISA close* block.

Error out: Connect this pin to the error pin of *VISA close*.

Error in: Connect this pin to the error out pin of *VISA configure serial port*.

VISA close: Closes a device session or an event object that is specified by the VISA resource name (Figure 4.16).

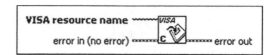

FIGURE 4.16
VISA close.

5

LabVIEW Interfacing with the Proteus Simulation Software

Before implementing the designed system on the actual hardware and interfacing it to the LabVIEW, it is preferred to check the Proteus model with the LabVIEW. If there would be any error then it can be solved at the software level only. To interface LabVIEW with the Proteus software, a virtual serial port emulator (VSPE) needs to be installed. VSPE creates the virtual bridge between the serial port of Proteus software with the LabVIEW simulator.

The system is first designed on the Proteus simulation and then interfaced with LabVIEW through VSPE to check its feasibility.

5.1 Virtual Serial Port Emulator

To set up the interfacing, install the VSPE software and open the window. Click on device and create as shown in Figure 5.1. A pop will appear and select *"pair"* as a device type and click next, as shown in Figure 5.2. Then assign the COM of COMPIM of Proteus and COM in LABVIEW as shown in Figure 5.3. Then click the add button on the software; it will pair and show the status in the device manager. Figure 5.4 shows the COMPIM configuration in the Proteus to interface with the LABVIEW. Figure 5.5 shows the COMPIM and Arduino connection in the Proteus.

The COMPIM model is a physical interface model (PIM) of a serial port. It receives serial data (9600-8-N-1) in buffer and transmits to the circuit as an input signal. The data from the CPU or UART appears at the PC's physical COM port.

Figure 5.1 shows the VSPE window that opens on clicking the software icon.

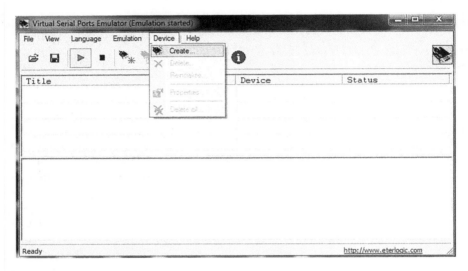

FIGURE 5.1
VSPE window.

Figure 5.2 shows the VSPE window for pairing the two virtual ports.

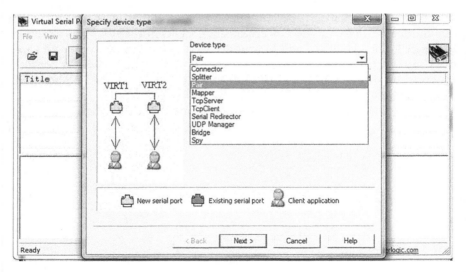

FIGURE 5.2
VSPE window for pairing two virtual ports.

Figure 5.3 shows how to assign COM ports and then finish the pairing process.

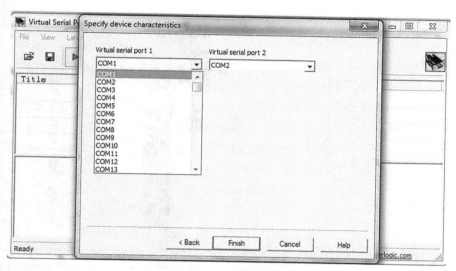

FIGURE 5.3
VSPE window for assigning COM port number.

Figure 5.4 shows the paired COM ports in VSPE. Let us assume that COM1 is with the Proteus model and COM2 is with LabVIEW.

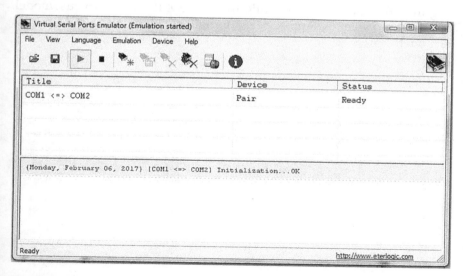

FIGURE 5.4
VSPE window showing paired COM ports.

Figure 5.5 shows how to interface COMPIM with the controller in Proteus model.

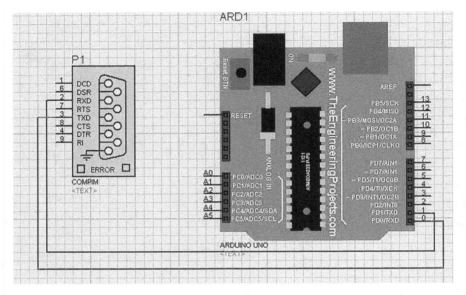

FIGURE 5.5
COMPIM and Arduino connection in Proteus.

Figure 5.6 shows how to configure the COMPIM in Proteus model. The edit component POP-up appears on right clicking on the COM port in the model.

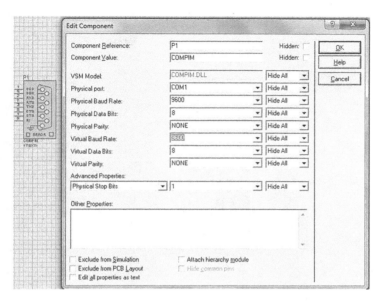

FIGURE 5.6
Configure the COMPIM in Proteus to interface with LabVIEW.

Figure 5.7 shows the Proteus simulation model displaying the sensor value on the virtual terminal after COM port pairing with VSPE. Figure 5.8 is pictorial representation of the complete process of interfacing Proteus Simulation Software with LabVIEW.

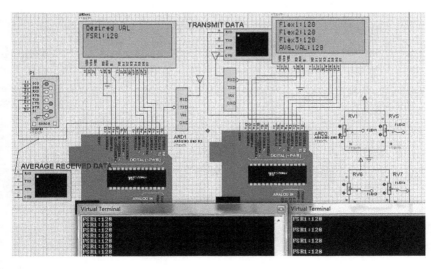

FIGURE 5.7
Proteus simulation model showing sensor value at virtual terminal.

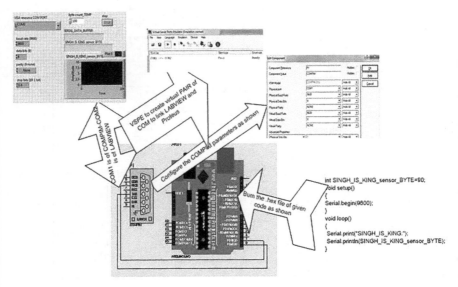

FIGURE 5.8
Arduino interfacing with the LabVIEW.

Section II

Arduino and I/O Devices

6

Arduino and Display Devices

This chapter describes the Arduino interfacing with the display devices such as the light emitting diode (LED) and the liquid crystal display (LCD). The working of the devices is discussed with the help of an interfacing circuit, program, and Proteus simulation models.

6.1 Arduino and Light Emitting Diode

A LED is a device, which can be used to indicate any condition or to display normal or warning conditions. LED has two terminals: (1) anode and (2) cathode. LED are available in the market in different colors. Figure 6.1 shows the snapshot of LED.

Different colors can be used to represent different conditions. The color of LED is due to compounds that emit light in specific regions of the visible light spectrum. Table 6.1 shows the characteristics of a typical LED.

To study the operation of LED, following components are required as given in Table 6.2.

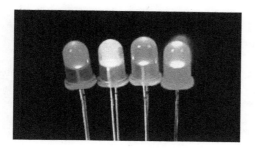

FIGURE 6.1
Light emitting diode.

TABLE 6.1

LED Characteristics

S.No.	Semiconductor Material	Wavelength	Color	V_F@20 mA
1	GaAs	850–940 nm	Infrared	1.2 V
2	GaAsP	630–660 nm	Red	1.8 V
3	GaAsP	605–620 nm	Amber	2.0 V
4	GaAsP:N	585–595 nm	Yellow	2.2 V
5	AlGaP	550–570 nm	Green	3.5 V
6	SiC	430–505 nm	Blue	3.5 V
7	GaInN	450 nm	White	4.0 V

Source: http://www.electronics-tutorials.ws/diode/diode_8.html

TABLE 6.2

Components List to Study the Working of LED

Component/Specification	Quantity
Power supply/+5 V/500 mA	1
Arduino Uno	1
LED	1
Connecting wires (M–M, M–F, F–F)	20 each
Zero-size PCB or bread board or a designed PCB	1

6.1.1 Circuit Diagram

LED connection

- Arduino digital pin 7- LED

Figure 6.2 shows the circuit diagram of the Arduino interfacing with the LED.

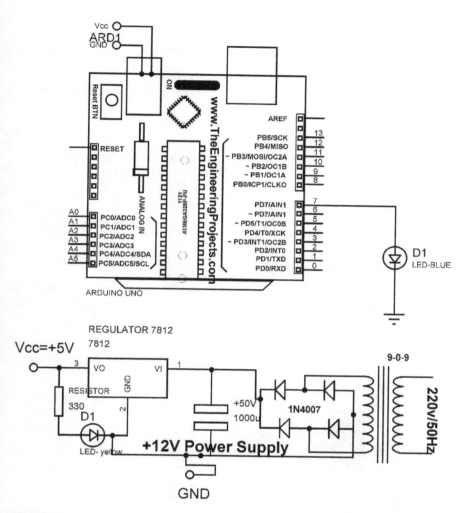

FIGURE 6.2
Circuit diagram to read LED.

6.1.2 Program

```
int LED_PIN=7;
void setup()
{
  pinMode(LED_PIN, OUTPUT);// initialize pin 7 as output pin
}

void loop()
{
 digitalWrite(LED_PIN, HIGH);      // Make pin 7 HIGH
 delay(1000);                       // 1000 mS delay
 digitalWrite(LED_PIN, LOW);      // Make pin 7 HIGH
 delay(1000);                       // 1000 mS delay
}
```

6.1.3 Proteus Simulation Model

Figure 6.3 shows the Proteus simulation model for interfacing the LED with the Arduino. Load the program in Arduino and check the working of the circuit. When simulation occurs in "run" condition, then it holds the "on" position and LED will glow.

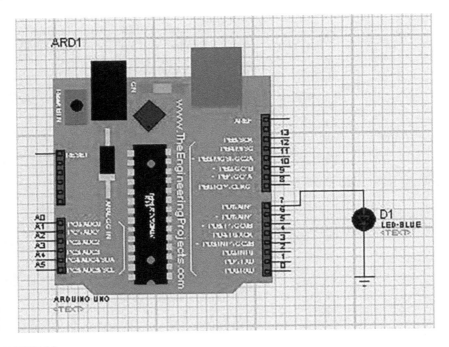

FIGURE 6.3
Proteus simulation model to read LED.

6.2 Arduino and Liquid Crystal Display

LCD is a display module. A 16 × 2 LCD display is commonly used as a display device in various circuits. This module is preferred over seven segments as they are economical, easily programmable, and have no limitation of displaying special and even custom characters.

A 16 × 2 LCD means that it can display 16 characters per row and there are two rows. In this LCD, each character is displayed in a 5 × 7 pixel matrix. This LCD has two registers: (1) command and (2) data.

The command register stores the instructions that are given to the LCD. An instruction is given to the LCD to do a predefined task such as initializing it, clearing its screen, setting the cursor position, and controlling the display, and so on. The data register stores the data to be displayed on the LCD (Figure 6.4).

A 20 × 4 LCD means that it can display 20 characters per row and there are 4 rows. In this LCD, each character is displayed in a 5 × 7 pixel matrix. The pin description is same as LCD (16 × 2). 20 × 4 LCD is considered to demonstrate the interfacing of LCD with the Arduino (Figure 6.5 and Table 6.3).

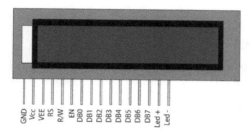

FIGURE 6.4
Liquid crystal display (16 × 2).

FIGURE 6.5
Liquid crystal display (20 × 4).

TABLE 6.3

Pin Description of a LCD

Pin	Description
1. (Ground)	Ground (0 V)
2. (V_{CC})	Power supply (5 V)
3. (V_{EE})	Contrast adjustment with a variable resistor
4. (Register select)	Selects command register when low and selects data register when high
5. (Read/write)	Low to write to the register and high to read the register
6. (Enable)	Send data to the data line when high to low pulse is given
7. (DB0)	8 bit data lines
8. (DB1)	
9. (DB2)	
10. (DB3)	
11. (DB4)	
12. (DB5)	
13. (DB6)	
14. (DB7)	
15. (LED+)	Backlight Vcc (5 V)
16. (LED−)	Backlight ground (0 V)

6.2.1 Circuit Diagram

Connect the Vcc of Arduino with +5 V and GND pin to ground of power supply.

LCD connection
- Arduino digital pin 13-RS pin(4) of LCD
- Arduino digital pin GND-RW pin(5) of LCD
- Arduino digital pin 12-E pin(6) of LCD
- Arduino digital pin 11-D4 pin(11) of LCD
- Arduino digital pin 10-D5 pin(12) of LCD
- Arduino digital pin 9-D6 pin(13) of LCD
- Arduino digital pin 8-D7 pin(14) of LCD

Figure 6.6 shows the circuit diagram of the Arduino interfacing with the LCD.

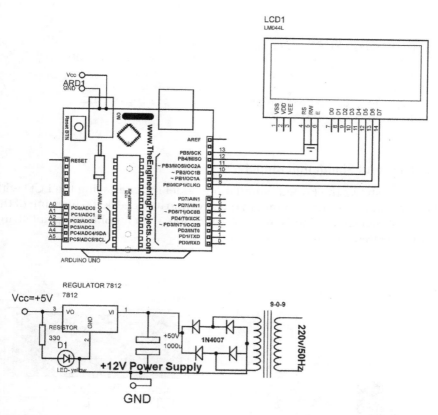

FIGURE 6.6
Circuit diagram to read LCD.

6.2.2 Program

```
#include <LiquidCrystal.h>
LiquidCrystal lcd(13, 12, 11, 10,9,8);

void setup()
{
  lcd.begin(20, 4);// Initialize LCD
  lcd.print("WELCOME_UPES");// Print string on LCD
  delay(2000);// Delay 2000mS
  lcd.clear();
}
```

```
void loop()
{
  lcd.setCursor(0, 1);// set cursor of LCD
  lcd.print("EIC Department");// Print string on LCD
  delay(2000);// Delay 2000mS
  lcd.setCursor(0, 2);// set cursor of LCD
  lcd.print("UPES Dehradun");// Print string on LCD
  delay(2000);// Delay 2000mS
}
```

6.2.3 Proteus Simulation Model

Figure 6.7 shows the Proteus simulation model for interfacing the LCD with the Arduino. Load the program in the Arduino and check the working of the circuit. When simulation is in "run" condition, then it holds the print string "UPES Dehradun" on the LCD.

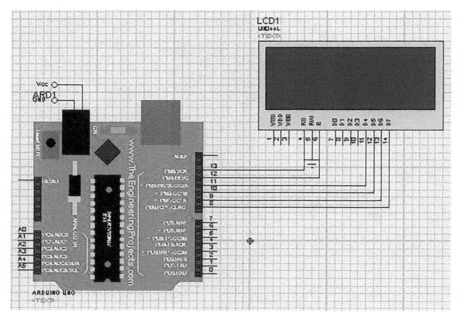

FIGURE 6.7
Proteus simulation model to read LCD.

7

Arduino and Digital Input/Output Devices

This chapter describes the Arduino interfacing with digital I/O devices such as a push button, fire sensor, passive infrared (PIR) sensor, and an alcohol sensor. The working of the devices is discussed with the help of interfacing circuit, program and proteus simulation models.

7.1 Push Button and Light Emitting Diode/Liquid Crystal Display

A push button is a switch mechanism for controlling some aspect of a machine or a process. Buttons are usually made of plastic or metal. The surface is flat so that it can be easily pressed. Buttons require a spring to return to their unpushed state. These are commonly used in calculators, telephones, and other electronic devices. Figure 7.1 shows the snapshot of the two terminal push buttons.

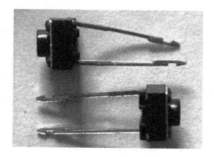

FIGURE 7.1
Push buttons.

Following components are required to study the operation of a push button (Table 7.1).

TABLE.7.1

Components List to Study the Working of a Push Button

Component/Specification	Quantity
Power supply/+12 V/1 A	1
Arduino Uno	1
Push button	1
RED LED	1
BLUE LED	1
LCD (20 * 4)	1
LCD patch	1
Connecting wires (M–M, M–F, F–F)	20 each
Zero-size PCB or bread board or a designed PCB	1

Figure 7.2 shows the block diagram to study the working of a push button with Arduino, which comprises Arduino, power supply, liquid crystal display (LCD), light emitting diode (LED), and a push button. LCD is connected to display the required content that corresponds to the action taken. LEDs are connected to show which action is taken, for example, if the button is not pressed "blue" LED will glow and if it is pressed then "red" LED will glow.

The programmer can read a push button by two methods: (1) digital low and (2) digital high.

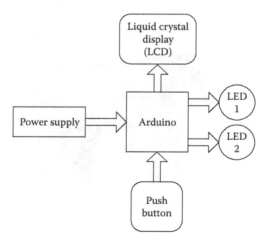

FIGURE 7.2
Block diagram for Arduino interfacing with the push button.

7.2 Push Button–Digital "LOW"

When the push button is unpressed, there is no connection between the two terminals of the push button, so the pin is connected to the ground and it is read as LOW. The red LED will glow if the switch is pressed or else the blue LED will glow.

7.2.1 Circuit Diagram

To read the push button as digital low, connect the components as follows:

Push button connection
- Arduino GND-push button one terminal
- Arduino digital pin 5-push button other terminal

LCD connection
- Arduino digital pin 13-RS pin(4) of LCD
- Arduino digital pin GND-RW pin(5) of LCD
- Arduino digital pin 12-E pin(6) of LCD
- Arduino digital pin 11-D4 pin(11) of LCD
- Arduino digital pin 10-D5 pin(12) of LCD
- Arduino digital pin 9-D6 pin(13) of LCD
- Arduino digital pin 8-D7 pin(14) of LCD

LED connection
- Arduino digital pin 3-RED-LED
- Arduino digital pin 2-BLUE-LED

Figure 7.3 shows the circuit diagram interfacing the push button with the Arduino.

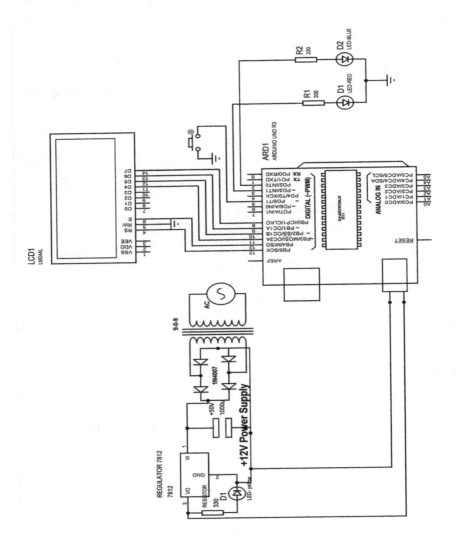

FIGURE 7.3
Circuit diagram to read the push button as digital LOW.

7.2.2 Program

```
#include <LiquidCrystal.h>
LiquidCrystal lcd(13, 12, 11, 10, 9, 8);// initialize the LCD
Library w.r t. RS,E,D4,D5,D6,D7
int BUTTON_LOW=5;
int RED_LED=3;
int BLUE_LED=2;
void setup()
{
pinMode(BUTTON_LOW, INPUT_PULLUP);//configure pin5 as an input
and enable the internal pull-up resistor
pinMode(RED_LED,OUTPUT);//
configure pin4 as outputpinMode(BLUE_LED,OUTPUT);//
configure pin3 as output
 lcd.begin(20, 4);// set the type of LCD by number of
 columns and rows
 lcd.setCursor(0, 0);// set cursor to column0 and row1
 lcd.print("DIGTAL LOW BUTTON ");// Print a message to
 the LCD.
 lcd.setCursor(0, 1);// set cursor to column0 and row1
 lcd.print("READ SYSTEM.......");// Print a message to the LCD.
 delay(1000);
}

void loop()
 {
   int BUTTON_LOW_READ = digitalRead(BUTTON_LOW);//read the
   pushbutton value into a variable
    if (BUTTON_LOW_READ == LOW)// Read PIN 5 as LOW PIN
   {
     lcd.clear();
     lcd.setCursor(0, 2);// set cursor to column0 and row2
     lcd.print("BUTTON_PRESSED ");// Print a message to the LCD.
     digitalWrite(RED_LED, HIGH);//High PIN3
     digitalWrite(BLUE_LED, LOW);// Low PIN2
     delay(20);
   }
   else //otherwise
   {
     lcd.clear();
     lcd.setCursor(0, 2);// set cursor to column0 and row2
     lcd.print("BUTTON_NOT_PRESSED ");// Print a message to
     the LCD.
     digitalWrite(BLUE_LED, HIGH);//High PIN2
     digitalWrite(RED_LED,LOW);//Low PIN3
     delay(20);
   }
 }
}
```

7.2.3 Proteus Simulation Model

Figures 7.4 and 7.5 show the Proteus simulation model for interfacing the push button with the Arduino. Load the program in Arduino and check for the working of the circuit. When button is pressed then it holds the "on" position and red LED will glow.

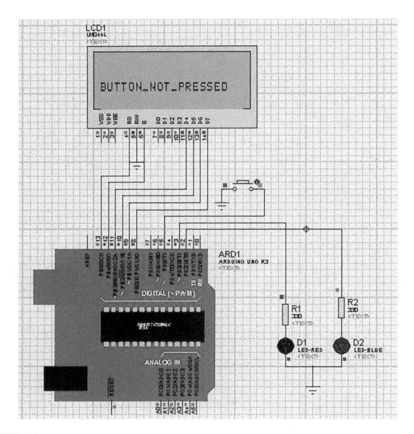

FIGURE 7.4
Proteus simulation model showing that the push button is not pressed.

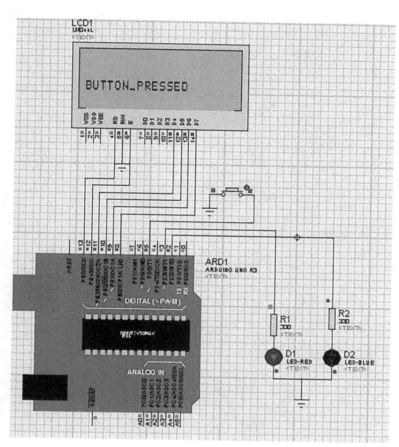

FIGURE 7.5
Proteus simulation model showing that the push button is pressed.

7.3 Push Button–Digital "HIGH"

When the button is closed (pressed), it makes a connection between the two terminals connecting the pin to 5 V, and it is read as HIGH. The circuit can be connected in a opposite way when a pull-up resistor keeps the input HIGH and goes LOW when the button is pressed; in this case, the behavior of LED will be opposite, it will be normally ON and become OFF when the switch is pressed. Figure 7.6 shows the circuit diagram to read the push button as digital "HIGH."

Components list is same as provided in Table 7.1.

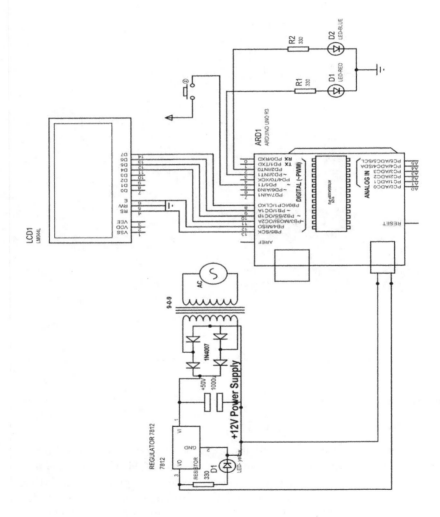

FIGURE 7.6
Circuit diagram to read the push button as digital "HIGH."

7.3.1 Program

```
#include <LiquidCrystal.h>
LiquidCrystal lcd(13, 12, 11, 10, 9, 8);// initialize the LCD
Library w.r t. RS,E,D4,D5,D6,D7
int BUTTON_HIGH=5;
int RED_LED=3;
int BLUE_LED=2;
void setup()
{
  pinMode(BUTTON_HIGH, INPUT);//configure pin5 as an input and
  enable the internal pull-up resistor
  pinMode(RED_LED,OUTPUT);//configure pin4 as output
  pinMode(BLUE_LED,OUTPUT);//configure pin3 as output
  lcd.begin(20, 4);// set up the LCD's number of columns
  and rows
  lcd.setCursor(0, 0);// set cursor to column0 and row1
  lcd.print("DIGTAL HIGH BUTTON ");// Print a message to
  the LCD.
  lcd.setCursor(0, 1);// set cursor to column0 and row1
  lcd.print("READ SYSTEM.......");// Print a message to the LCD.
  delay(1000);
}

void loop()
{

  int BUTTON_HIGH_READ = digitalRead(BUTTON_HIGH);//read the
  pushbutton value into a variable
    if (BUTTON_HIGH_READ == HIGH)// Read PIN 5 as LOW PIN
    {
      lcd.clear();
      lcd.setCursor(0, 2);// set cursor to column0 and row2
      lcd.print("BUTTON_PRESSED ");// Print a message to the LCD.
      digitalWrite(RED_LED, HIGH);//High PIN3
      digitalWrite(BLUE_LED, LOW);// Low PIN2
      delay(20);
    }
  else //otherwise
    {
      lcd.clear();
      lcd.setCursor(0, 2);// set cursor to column0 and row2
      lcd.print("BUTTON_NOT_PRESSED ");// Print a message to
      the LCD.
      digitalWrite(BLUE_LED, HIGH);//High PIN2
      digitalWrite(RED_LED,LOW);//Low PIN3
      delay(20);
    }
}
```

7.3.2 Proteus Simulation Model

Figures 7.7 and 7.8 show the Proteus simulation model for the push button with Arduino. Load the program in the Arduino and check the working of the circuit. When the button is pressed then it holds the "on" position and Red LED will glow.

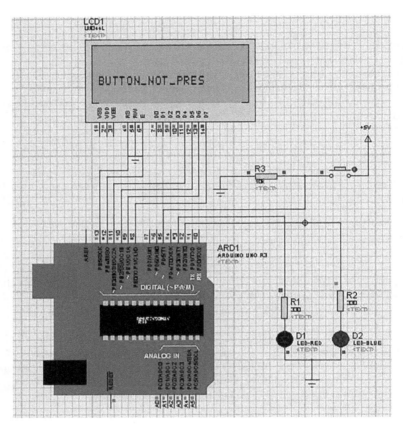

FIGURE 7.7
Proteus simulation model to read the push button as digital "HIGH" (button not pressed).

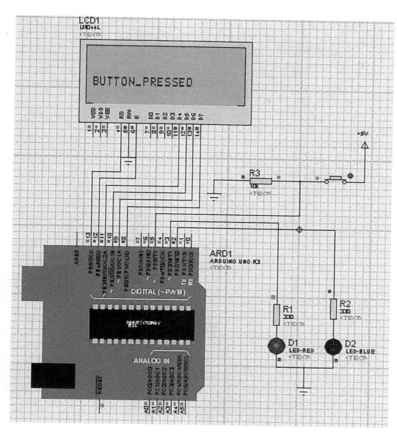

FIGURE 7.8
Proteus simulation model to read the push button as digital "HIGH" (button pressed).

7.4 Fire Sensor and Light Emitting Diode/Liquid Crystal Display

The flame sensor is a digital sensor, which detects fire. To demonstrate its working, fire sensor from "Robosoft systems" is considered. The module comprises of an IR sensor, OP–Amp circuitry, potentiometer, and a LED as an indicator (Table 7.2). RED LED glows when the sensor detects the fire.

Arduino-Based Embedded Systems

TABLE 7.2

Components List to Study the Working of a Fire Sensor

Component/Specification	Quantity
Power supply/+12 V/1 A, +5 V/500 mA	1
Arduino Uno	1
Fire sensor	1
Push button	1
RED LED	1
BLUE LED	1
LCD (20 * 4)	1
LCD patch	1
Connecting wires (M–M, M–F, F–F)	20 each
Zero-size PCB or bread board or a designed PCB	1

FIGURE 7.9
Fire detector sensor module (Robosoft systems).

The sensitivity of sensor can be adjusted with the help of a potentiometer. The detecting range of this module is 1–2 m. It is easy to mount on any system due to its light weight of about 5 g. It gives a high output on fire detection.

Figure 7.9 shows the snapshot for the fire detection sensor.

Figure 7.10 shows the block diagram of the system to study the working of fire sensor. It comprises of the Arduino, power supply, LEDs, and fire sensor. LEDs are connected to indicate the action taken; for example, if the fire sensor senses the fire then "red" LED will glow else the "blue" LED will glow.

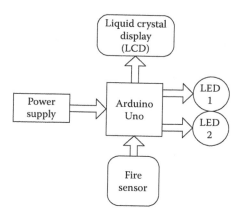

FIGURE 7.10
Block diagram to study the fire sensor.

7.4.1 Circuit Diagram

FLAME sensor connection

- Arduino GND-Module GND
- Arduino +5 V-Module +
- Arduino digital pin 2-Module digital out pin

LCD connection

- Arduino digital pin 13-RS pin(4) of LCD
- Arduino digital pin GND-RW pin(5) of LCD
- Arduino digital pin 12-E pin(6) of LCD
- Arduino digital pin 11-D4 pin(11) of LCD
- Arduino digital pin 10-D5 pin(12) of LCD
- Arduino digital pin 9-D6 pin(13) of LCD
- Arduino digital pin 8-D7 pin(14) of LCD

LED connection

- Arduino digital pin 7-RED-LED
- Arduino digital pin 6-BLUE-LED

Figure 7.11 shows the circuit diagram to study the working of the fire sensor. Connect all the components as described in Section 7.4.1. Red LED will glow, when it will detect fire or else Blue LED will glow.

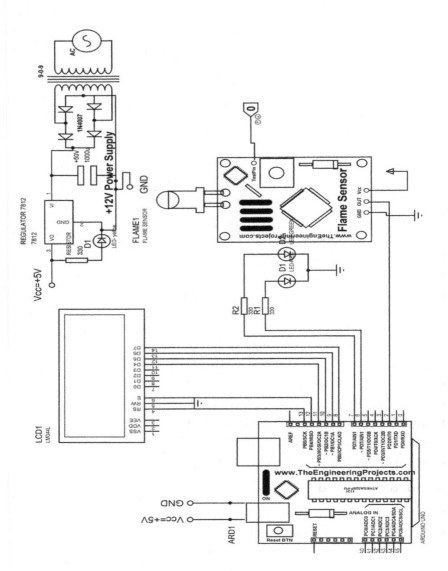

FIGURE 7.11
Circuit diagram to study the working of the fire sensor.

7.4.2 Program

```
#include <LiquidCrystal.h>
LiquidCrystal lcd(13, 12, 11, 10, 9, 8);// initialize the
LCD Library w.r t. RS,E,D4,D5,D6,D7
int FIRE_SENSOR_LOW=2;
int RED_LED=7;
int GREEN_LED=6;
void setup()
{
  pinMode(FIRE_SENSOR_LOW, INPUT_PULLUP);//configure pin5 as
  an input and enable the internal pull-up resistor
  pinMode(RED_LED,OUTPUT);//configure pin4 as output
  pinMode(GREEN_LED,OUTPUT);//configure pin3 as output
  lcd.begin(20, 4);// set up the LCD's number of columns and rows
  lcd.setCursor(0, 0);// set cursor to column0 and row1
  lcd.print("FIRE SENSOR BASED  ");// Print a message to the LCD.
  lcd.setCursor(0, 1);// set cursor to column0 and row1
  lcd.print(" FIRE DETECTION SYSTEM...");// Print a message to
  the LCD.
  delay(1000);
}

void loop()
{

  int FIRE_SENSOR_LOW_READ = digitalRead(FIRE_SENSOR_LOW);//
  read the pushbutton value into a variable
    if (FIRE_SENSOR_LOW_READ == LOW)// Read PIN 5 as LOW PIN
    {
    lcd.clear();
    lcd.setCursor(0, 2);// set cursor to column0 and row2
    lcd.print("FIRE DETECTED ");// Print a message to the LCD.
    digitalWrite(RED_LED, HIGH);//High PIN3
    digitalWrite(GREEN_LED, LOW);// Low PIN2
    delay(20);
    }
  else //otherwise
    {
    lcd.clear();
    lcd.setCursor(0, 2);// set cursor to column0 and row2
    lcd.print("FIRE NOT DETECTED ");// Print a message to the LCD.
    digitalWrite(GREEN_LED, HIGH);//High PIN2
    digitalWrite(RED_LED,LOW);//Low PIN3
    delay(20);
    }
}
```

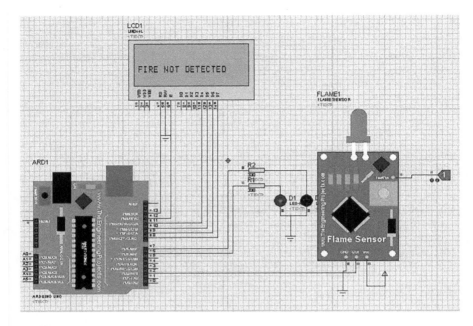

FIGURE 7.12
Proteus simulation model to study the fire sensor (fire not detected).

7.4.3 Proteus Simulation Model

Figures 7.12 and 7.13 show the Proteus simulation model for the system. Connect all the components as described in Section 7.3.2. In addition to it, a "logic" needs to be connected with the sensor. As this is a virtual environment, so to make its input as LOW or HIGH, a logic is connected to it, which can be changed manually to check the working of the sensor.

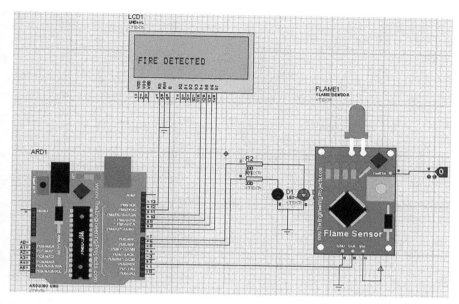

FIGURE 7.13
Proteus simulation model to study the fire sensor (fire detected).

7.5 Passive Infrared Sensor and Light Emitting Diode/Liquid Crystal Display

PIR sensor detects the motion. PIR sensor from sunrom-1133 is considered for demonstration. It has a Fresnel lens and a motion detection circuit, which is suitable for a wide range of supply voltages with low current drain. It gives output as standard TTL active low signal indicated by an onboard LED. It has a range of 6 m and can be used to access control systems and burglar alarms. It has three terminals—(1) power, (2) GND, and (3) an output.

Figure 7.14 shows the snapshot of the PIR sensor.

Figure 7.15 shows the block diagram of the system to study the working of the PIR sensor. It comprises the Arduino, power supply, LEDs, and the PIR sensor (Table 7.3). LEDs are connected to show, which input is given by the sensor, for example, if the sensor detects any motion then "red" LED will glow or else "blue" LED will glow.

FIGURE 7.14
PIR sensor (sunrom-1133).

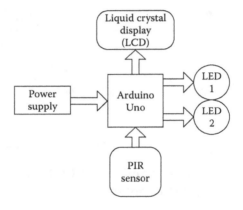

FIGURE 7.15
Block diagram to study the working of a PIR sensor.

TABLE 7.3

Components List to Study the Working of a PIR Sensor

Component/Specification	Quantity
Power supply/+12 V/1 A, +5 V/500 mA	1
Arduino Uno	1
PIR sensor	1
Push button	1
RED LED	1
BLUE LED	1
LCD (20 * 4)	1
LCD patch	1
Connecting wires (M–M, M–F, F–F)	20 each
Zero-size PCB or bread board or a designed PCB	1

7.5.1 Circuit Diagram

PIR sensor connection

- Arduino GND-Module GND
- Arduino +5 V-Module +
- Arduino digital pin 2-Module digital out pin

LCD connection

- Arduino digital pin 13-RS pin(4) of LCD
- Arduino digital pin GND-RW pin(5) of LCD
- Arduino digital pin 12-E pin(6) of LCD
- Arduino digital pin 11-D4 pin(11) of LCD
- Arduino digital pin 10-D5 pin(12) of LCD
- Arduino digital pin 9-D6 pin(13) of LCD
- Arduino digital pin 8-D7 pin(14) of LCD

LED connection

- Arduino digital pin 7-RED-LED
- Arduino digital pin 6-BLUE-LED

Figure 7.16 shows the circuit diagram of the system to study the working fire sensor, comprising Arduino, power supply, LEDs, and the fire sensor.

7.5.2 Program

```
#include <LiquidCrystal.h>
LiquidCrystal lcd(13, 12, 11, 10, 9, 8);// initialize
the LCD Library w.r t. RS,E,D4,D5,D6,D7
int PIR_SENSOR_LOW=5;
int RED_LED=7;
int BLUE_LED=6;
void setup()
{
  pinMode(PIR_SENSOR_LOW, INPUT_PULLUP);//configure pin5 as
  an input and enable the internal pull-up resistor
  pinMode(RED_LED,OUTPUT);//configure pin4 as output
  pinMode(BLUE_LED,OUTPUT);//configure pin3 as output
```

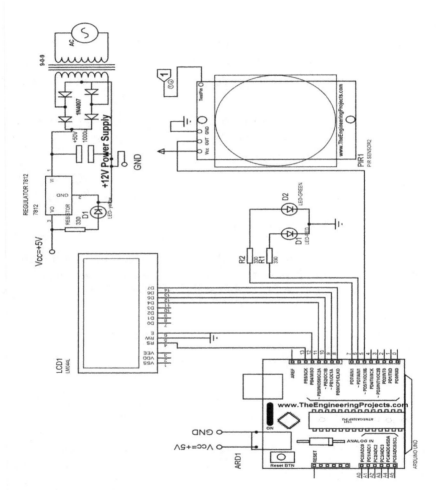

FIGURE 7.16
Circuit diagram to study the working of a PIR sensor.

```
lcd.begin(20, 4);// set up the LCD's number of columns
and rows
lcd.setCursor(0, 0);// set cursor to column0 and row1
lcd.print("MOTION SENSOR BASED");// Print a message to the LCD.
lcd.setCursor(0, 1);// set cursor to column0 and row1
lcd.print("MOTION DETECTION ");// Print a message to the LCD.
lcd.setCursor(0, 2);// set cursor to column0 and row2
lcd.print("SYSTEM AT UPES");// Print a message to the LCD.
delay(1000);
}

void loop()
{

  int PIR_SENSOR_LOW_READ = digitalRead(PIR_SENSOR_LOW);//
  read the pushbutton value into a variable
    if (PIR_SENSOR_LOW_READ == LOW)// Read PIN 5 as LOW PIN
    {
    lcd.clear();
    lcd.setCursor(0, 3);// set cursor to column0 and row2
    lcd.print("MOTION DETECTED ");// Print a message to the LCD.
    digitalWrite(RED_LED, HIGH);//High PIN3
    digitalWrite(BLUE_LED, LOW);// Low PIN2
    delay(20);
    }
  else //oterwise
  {
    lcd.clear();
    lcd.setCursor(0, 3);// set cursor to column0 and row3
    lcd.print("MOTION NOT DETECTED ");// Print a message to
    the LCD.
    digitalWrite(BLUE_LED, HIGH);//High PIN2
    digitalWrite(RED_LED,LOW);//Low PIN3
    delay(20);
  }
}
```

7.5.3 Proteus Simulation Model

Figures 7.17 and 7.18 show the Proteus simulation model for the PIR sensor. Connect all the components as described in Section 7.3.2. In addition to it, a "logic" needs to be connected with the sensor. As this is virtual environment, so to make its input as LOW or HIGH, a "logic" is connected to it, which can be changed manually to check the working of the sensor.

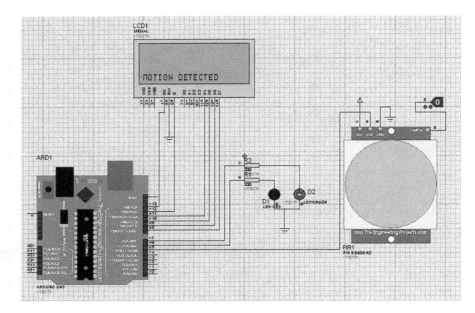

FIGURE 7.17
Proteus simulation model for a PIR sensor (motion detected).

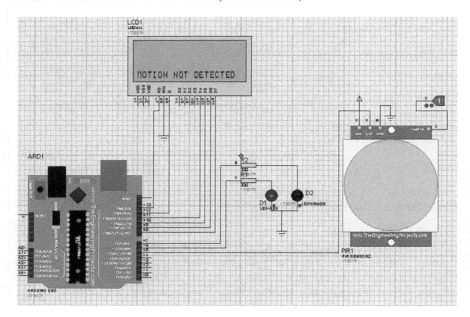

FIGURE 7.18
Proteus simulation model for a PIR sensor (motion not detected).

7.6 Alcohol Sensor and Light Emitting Diode/Liquid Crystal Display

Alcohol sensor is a semiconductor device, which detects the presence of alcohol. To demonstrate the working of an alcohol sensor, MQ3 sensor from sunrom is considered, which can detect alcohol with concentration ranging from 0.05 to 10 mg/L. It is made of a sensitive material SnO_2; it shows lower conductivity in clean air, and conductivity increase with an increase in the alcohol–gases concentration. This module gives output in both: (1) the analog forms and (2) the digital forms.

This alcohol sensor can be used to detect alcohol consumption by a person from his or her breath. Figure 7.19 shows the snapshot of an alcohol sensor. Table 7.4 describes the components required to study the alcohol sensor.

FIGURE 7.19
Alcohol sensor from sunrom.

TABLE 7.4

Components List to Study the Working of an Alcohol Sensor

Component/Specification	Quantity
Power supply/+12 V/1 A, +5 V/500 mA	1
Arduino Uno	1
Alcohol sensor	1
Push button	1
RED LED	1
BLUE LED	1
LCD (20 * 4)	1
LCD patch	1
Sensor patch	1
Connecting wires (M–M, M–F, F–F)	20 each
Zero-size PCB or bread board or a designed PCB	1

Figure 7.20 shows the block diagram of the system to study the working of an alcohol sensor. It comprises Arduino, power supply, LEDs, and an alcohol sensor.

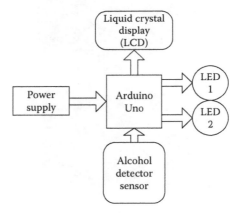

FIGURE 7.20
Block diagram to study the working of an alcohol sensor.

7.6.1 Circuit Diagram

Alcohol sensor connection with Arduino
- Arduino GND-Module GND
- Arduino +5 V-Module +
- Arduino digital pin 2-Module digital out pin

LCD connection
- Arduino digital pin 13-RS pin(4) of LCD
- Arduino digital pin GND-RW pin(5) of LCD
- Arduino digital pin 12-E pin(6) of LCD
- Arduino digital pin 11-D4 pin(11) of LCD
- Arduino digital pin 10-D5 pin(12) of LCD
- Arduino digital pin 9-D6 pin(13) of LCD
- Arduino digital pin 8-D7 pin(14) of LCD

LED connection
- Arduino digital pin 5-RED-LED
- Arduino digital pin 4-BLUE-LED

Figure 7.21 shows the circuit diagram to study the alcohol sensor.

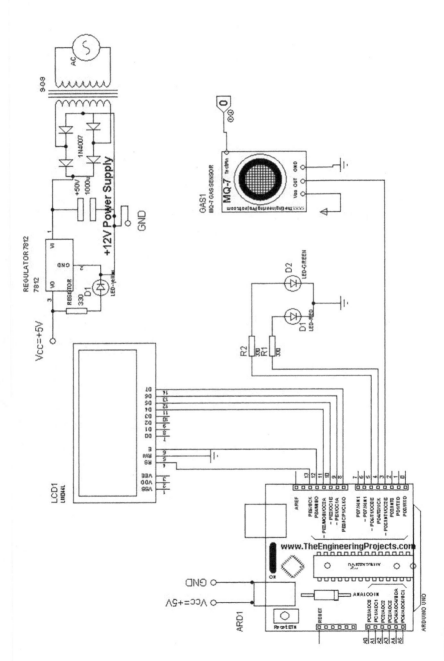

FIGURE 7.21

Circuit diagram to study an alcohol sensor.

7.6.2 Program

```
#include <LiquidCrystal.h>
LiquidCrystal lcd(13, 12, 11, 10, 9, 8);// initialize the
LCD Library w.r t. RS,E,D4,D5,D6,D7
int ALCOHOL_SENSOR_LOW=3;
int RED_LED=5;
int GREEN_LED=4;
void setup()
{
 pinMode(ALCOHOL_SENSOR_LOW, INPUT_PULLUP);//configure pin5 as
an input and enable the internal pull-up resistor
 pinMode(RED_LED,OUTPUT);//configure pin4 as output
 pinMode(GREEN_LED,OUTPUT);//configure pin3 as output
 lcd.begin(20, 4);// set the type of LCD as per number of
columns and rows
 lcd.setCursor(0, 0);// set cursor to column0 and row1
 lcd.print("ALCOHOL SENSOR BASED");// Print a message to the
LCD.
 lcd.setCursor(0, 1);// set cursor to column0 and row1
 lcd.print("ALCOHOL DETECTION ");// Print a message to
the LCD.
 lcd.setCursor(0, 2);// set cursor to column0 and row2
 lcd.print("SYSTEM AT UPES.....");// Print a message to the LCD.
 delay(1000);
}


void loop()
{

 int ALCOHOL_SENSOR_LOW_READ = digitalRead(ALCOHOL_SENSOR_
LOW);//read the pushbutton value into a variable
  if (ALCOHOL_SENSOR_LOW_READ == LOW)// Read PIN 5 as LOW PIN
 {
 lcd.clear();
 lcd.setCursor(0, 3);// set cursor to column0 and row2
 lcd.print("ALCOHOL DETECTED ");// Print a message to
the LCD.
 digitalWrite(RED_LED, HIGH);//High PIN3
 digitalWrite(GREEN_LED, LOW);// Low PIN2
 delay(20);
 }
```

```
else //oterwise
{
  lcd.clear();
  lcd.setCursor(0, 3);// set cursor to column0 and row3
  lcd.print("ALCOHOL NOT DETECTED ");// Print a message to
  the LCD.
  digitalWrite(GREEN_LED, HIGH);//High PIN2
  digitalWrite(RED_LED,LOW);//Low PIN3
  delay(20);
 }
}
```

7.6.3 Proteus Simulation Model

Figures 7.22 and 7.23 show the Proteus simulation model for the system. Connect all the components as described in Section 7.4.2. In addition to it, a "logic" needs to be connected with the sensor. As this is a virtual environment, so to make its input as LOW or HIGH, a logic is connected to it, which can be changed manually to check the working of the sensor.

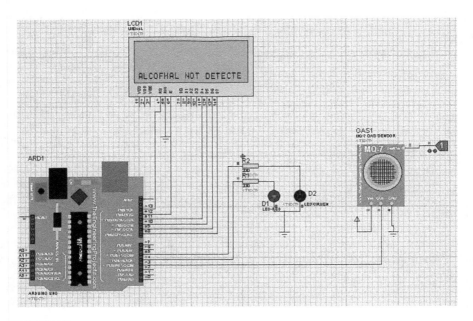

FIGURE 7.22
Proteus simulation model for an alcohol sensor (alcohol not detected).

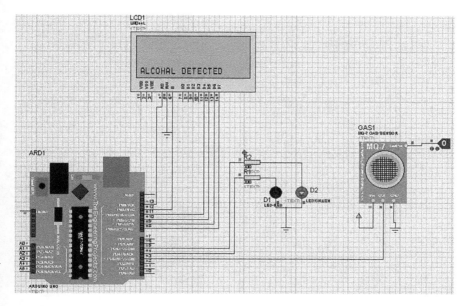

FIGURE 7.23
Proteus simulation model for an ultrasonic sensor (alcohol detected).

8

Arduino and Analog Devices

This chapter describes the Arduino interfacing with the analog devices such as ultrasonic sensor, passive infrared (PIR) sensor, and an alcohol sensor. The working of the devices is discussed with the help of the interfacing circuit, program, and proteus simulation models.

8.1 Ultrasonic Sensor and Liquid Crystal Display

Ultrasonic sensor measures the distance of an object from the sensor. It has a small size and is easy to use. It measures the distance with the help of sound waves. It sends out a sound wave at a specific frequency, which when strike through the object reflects back to its origin. By recording the elapsed time between the sound wave being generated and the sound wave bouncing back, it is possible to calculate the distance between the sonar sensor and the object.

Figure 8.1 shows the block diagram to study the working of ultrasonic sensor, comprising Arduino, ultrasonic sensor, and liquid crystal display (LCD) (Table 8.1).

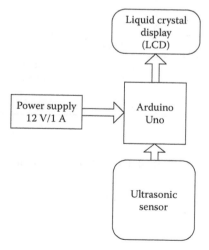

FIGURE 8.1
Block diagram to study an ultrasonic sensor.

TABLE 8.1

Component List to Study the Working of an Ultrasonic Sensor

Component/Specification	Quantity
Power supply/+12 V/1 A, +5 V/500 mA	1
Arduino Uno	1
Ultrasonic sensor	1
LCD (20 * 4)	1
LCD patch	1
Sensor patch	1
Connecting wires (M–M, M–F, F–F)	20 each
Zero-size PCB or bread board or a designed PCB	1

Ultrasonic sensor can be connected to the controller in two modes-serial out and PMW out.

8.2 Ultrasonic Sensor—Serial Out

Model no. 1166 for ultrasonic sensor from Sunrom technologies gives serial output. Its operating range is 10–400 cm with accuracy of ±1 cm. It gives serial output at a baud rate of 9600 bps. It gives data in a packet format of 9 bytes as **xxx.xxxcm <CR>**.

Packet Format								
ASCII (0–9)	ASCII (0–9)	ASCII (0–9)	Fixed decimal (.)	ASCII (0–9)	ASCII (0–9)	ASCII (0–9)	Fixed ASCII character (c)	Fixed ASCII character (m)

Figure 8.2 shows the snapshot of an ultrasonic sensor from Sunrom with the Model no. 1166.

FIGURE 8.2
An ultrasonic sensor (Sunrom-1166).

8.2.1 Circuit Diagram

To read the ultrasonic sensor serial out, connect the components as follows:

Ultrasonic sensor connection in serial out mode
- Arduino GND-Module GND
- Arduino +5 V-Module +
- Arduino RX pin(0)-data out pin of sensor

LCD connection
- Arduino digital pin 13-RS pin(4) of LCD
- Arduino digital pin GND-RW pin(5) of LCD
- Arduino digital pin 12-E pin(6) of LCD
- Arduino digital pin 11-D4 pin(11) of LCD
- Arduino digital pin 10-D5 pin(12) of LCD
- Arduino digital pin 9-D6 pin(13) of LCD
- Arduino digital pin 8-D7 pin(14) of LCD

Figure 8.3 shows the circuit diagram interfacing the ultrasonic sensor with Arduino.

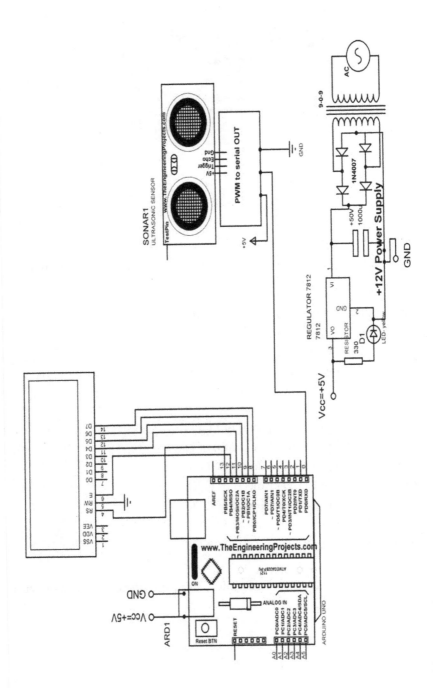

8.2.2 Program

```
#include <LiquidCrystal.h>
LiquidCrystal lcd(12, 11, 5, 4, 3, 2);
String inputString_Ultrasonic_serialout = "";  // a string
to hold incoming data
boolean stringComplete_Ultrasonic_serialout = false;
// whether the string is complete

void setup( ) {
 // initialize serial:
 Serial.begin(9600);
 lcd.begin(20, 4);
 inputString_Ultrasonic_serialout.reserve(200);
}

void loop()
{
 if (stringComplete_Ultrasonic_serialout)
 {
  lcd.clear();
  lcd.print(inputString_Ultrasonic_serialout);
  Serial.println(inputString_Ultrasonic_serialout);
  lcd.setCursor(0,3);
  lcd.print(inputString_Ultrasonic_serialout[0]);
  lcd.print(inputString_Ultrasonic_serialout[1]);
  lcd.print(inputString_Ultrasonic_serialout[2]);
  lcd.print(inputString_Ultrasonic_serialout[3]);
  lcd.print(inputString_Ultrasonic_serialout[4]);
  lcd.print(inputString_Ultrasonic_serialout[5]);

  if((inputString_Ultrasonic_serialout[1]>='3')&&
  (inputString_Ultrasonic_serialout[2]>='5'))
  {
  lcd.setCursor(0,2);
  lcd.print("WATER LEVEL OVER");

  }
 else
  {
  lcd.setCursor(0,2);
  lcd.print("WATER LEVEL OK");
  }
 inputString_Ultrasonic_serialout = "";
  stringComplete_Ultrasonic_serialout = false;
 }
}
```

```
void serialEvent()
{
  while (Serial.available())
  {
    char inChar = (char)Serial.read();
    inputString_Ultrasonic_serialout += inChar;
  if (inChar == 0x0D)
    {
    stringComplete_Ultrasonic_serialout = true;
    }
  }
}
```

8.2.3 Proteus Simulation Model

Figure 8.4 shows the Proteus simulation model for interfacing the ultrasonic sensor with the Arduino. Load the program in the Arduino and check the working of the circuit. To change the distance in the virtual environment, a potentiometer is connected.

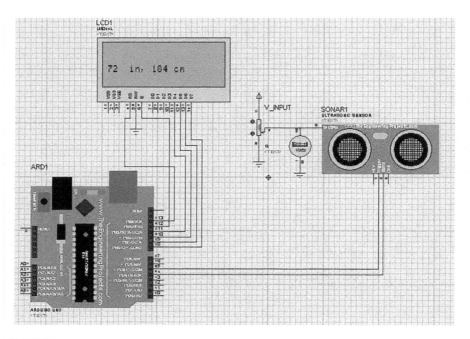

FIGURE 8.4
Proteus simulation model showing the working of an ultrasonic sensor.

8.3 Ultrasonic Sensor—PWM Out

Ultrasonic sensor with Sunrom model no. 3719 gives PWM output with two I/O pins. It can measure the distance ranging from 2 cm to 5 m. The sensor transmits an ultrasonic wave and produces output by measuring the echo pulse width. It operates on a +5 V DC. Figure 8.5 shows the ultrasonic sensor with PWM out.

FIGURE 8.5
An ultrasonic sensor (Sunrom-3719).

8.3.1 Circuit Diagram

PWM OUT

To read the ultrasonic PWM out, connect the components as follows:

Ultrasonic sensor PWM OUT connection
- Arduino GND-Module GND
- Arduino +5 V-Module +
- Arduino digital pin 5-Module digital trigger pin
- Arduino digital pin 4-Module digital echo pin

LCD connection
- Arduino digital pin 13-RS pin(4) of LCD
- Arduino digital pin GND-RW pin(5) of LCD
- Arduino digital pin 12-E pin(6) of LCD
- Arduino digital pin 11-D4 pin(11) of LCD
- Arduino digital pin 10-D5 pin(12) of LCD
- Arduino digital pin 9-D6 pin(13) of LCD
- Arduino digital pin 8-D7 pin(14) of LCD

Figure 8.6 shows the circuit diagram of an ultrasonic sensor in the PWM mode.

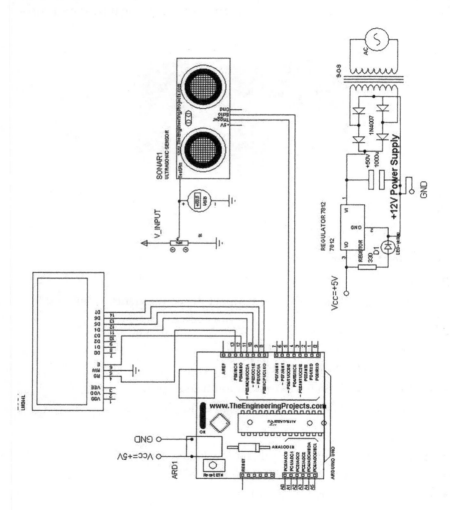

FIGURE 8.6
Circuit diagram to study the working of an ultrasonic sensor (PWM mode).

8.3.2 Program

8.3.2.1 *Ultrasonic Sensor—PWM OUT*

```
#include <LiquidCrystal.h>
LiquidCrystal lcd(13, 12, 11, 10, 9, 8); // initialize the LCD
pin RS,E,D4,D5,D6,D7
const int pingPin = 5; // Trigger Pin of Ultrasonic Sensor
const int echoPin = 4; // Echo Pin of Ultrasonic Sensor

void setup()
{
  lcd.begin(20, 4);
  lcd.setCursor(0, 0);
  lcd.print("Ultrasonic distance");
  lcd.setCursor(0, 1);
  lcd.print("System at UPES");
  delay(1000);
}

void loop()
{
  long duration, inches, cm;
  pinMode(pingPin, OUTPUT);
  digitalWrite(pingPin, LOW);
  delayMicroseconds(2);
  digitalWrite(pingPin, HIGH);
  delayMicroseconds(10);
  digitalWrite(pingPin, LOW);
  pinMode(echoPin, INPUT);
  duration = pulseIn(echoPin, HIGH);
  inches = microsecondsToInches(duration);
  cm = microsecondsToCentimeters(duration);
  lcd.clear();
  lcd.setCursor(0, 2);
  lcd.print(inches);
  lcd.setCursor(4, 2);
  lcd.print("in, ");
  lcd.setCursor(8, 2);
  lcd.print(cm);
  lcd.setCursor(12, 2);
  lcd.print("cm");
  Serial.print("in, ");
  Serial.print(cm);
  Serial.print("cm");
  Serial.println();*/
  delay(200);
}
```

```
long microsecondsToInches(long microseconds)
    {
    return microseconds / 74 / 2;
    }
long microsecondsToCentimeters(long microseconds)
    {
    return microseconds / 29 / 2;
    }
```

8.3.3 Proteus Simulation Model

Figure 8.7 shows the Proteus simulation model for interfacing the ultrasonic sensor with the Arduino. Load the program in Arduino and check the working of the circuit. To change the distance in the virtual environment, a potentiometer is connected.

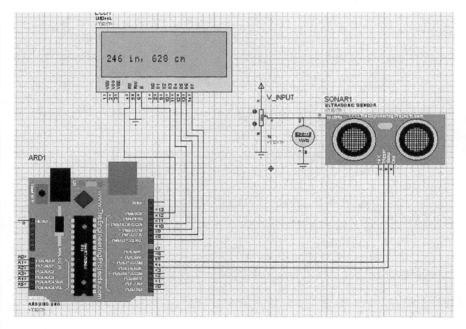

FIGURE 8.7
Proteus simulation model to study the working of an ultrasonic sensor (PWM mode).

8.4 Temperature Sensor and Liquid Crystal Display

A temperature sensor measures the temperature of surroundings according to the level of an electrical signal. The working of the sensor is based on the voltage difference. Whenever the temperature rises, the voltage also increases. The sensor records any change that occurs in the voltage between the transistor base and the emitter.

Figure 8.8 shows the block diagram of the system to measure the temperature and display it on the LCD. It comprises the controller that is Arduino Uno, power supply, LCD, and temperature sensor, which give an analog output (Table 8.2).

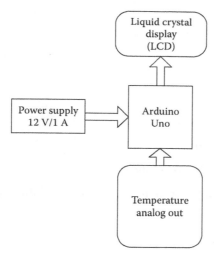

FIGURE 8.8
Block diagram to study the temperature sensor.

TABLE 8.2

Component List to Study the Working of Temperature Sensor

Component/Specification	Quantity
Power supply/+12 V/1 A, + 5 V/500 mA	1
Arduino Uno	1
Temperature sensor	1
LCD (20 * 4)	1
LCD patch	1
Sensor patch	1
Connecting wires (M–M, M–F, F–F)	20 each
Zero-size PCB or bread board or a designed PCB	1

8.5 Temperature Sensor-Analog Out

The LM35 series is an integrated circuit for temperature sensor. Corresponding to each degree Celsius (°C) it gives 10 mV as output. It can sense temperature in the range of −55°C−+150°C.

FIGURE 8.9
LM35.

8.5.1 Circuit Diagram

8.5.1.1 Temperature Sensor Analog Out

To read the temperature sensor analog out, connect the components as follows:

LM35 sensor connection
- Arduino GND-Module GND
- Arduino +5 V-Module +
- Arduino A0 pin-data out pin of sensor

LCD connection
- Arduino digital pin 13-RS pin(4) of LCD
- Arduino digital pin GND-RW pin(5) of LCD
- Arduino digital pin 12-E pin(6) of LCD
- Arduino digital pin 11-D4 pin(11) of LCD
- Arduino digital pin 10-D5 pin(12) of LCD
- Arduino digital pin 9-D6 pin(13) of LCD
- Arduino digital pin 8-D7 pin(14) of LCD

Figure 8.10 shows the circuit diagram for the temperature sensor interfacing with the Arduino.

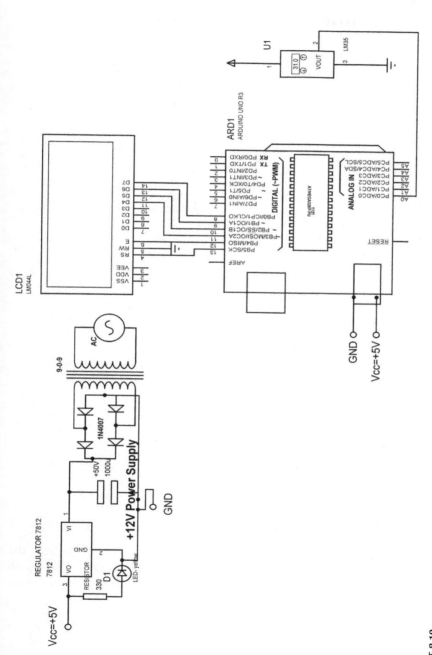

FIGURE 8.10

Circuit diagram to interface temperature sensor and Arduino.

8.5.2 Program

8.5.2.1 Program LM35—Analog Out

```
#include <LiquidCrystal.h>
LiquidCrystal lcd(13, 12, 11, 10,9, 8);// attach LCD pin
RS,E,D4,D5,D6,D7 to the given pins
int TEMP_sensor_Pin = A0;  // select the input pin for the
potentiometer
int TEMP_sensor_ADC_Value = 0;  // variable to store the
value coming from the sensor

void setup()
{

 lcd.begin(20, 4);// Initialise 20*4 LCD
 lcd.setCursor(0, 0);// set cursor of LCD at column0 and Row0
 lcd.print("Temperature monitoring");
 lcd.setCursor(0, 1);
 lcd.print("System at UPES.....");
 delay(1000);
 lcd.clear();
}
void loop()
{
 TEMP_sensor_ADC_Value = analogRead(TEMP_sensor_Pin);
 // read the value from the sensor
 float TEMP_sensor_Value_ACTUAL=TEMP_sensor_ADC_Value/2;
 lcd.setCursor(0,2);
 lcd.print("ADC LEVEL:");
 lcd.setCursor(11,2);
 lcd.print(TEMP_sensor_ADC_Value);
 lcd.setCursor(0,3);
 lcd.print("TEMP VALUE:");
 lcd.setCursor(12,3);
 lcd.print(TEMP_sensor_Value_ACTUAL);
}
```

8.5.3 Proteus Simulation Model

Figure 8.11 shows the Proteus simulation model for interfacing the temperature sensor with the Arduino. Load the program in the Arduino and check the working of the circuit. The LCD displays the ADC level of sensor and the corresponding temperature value.

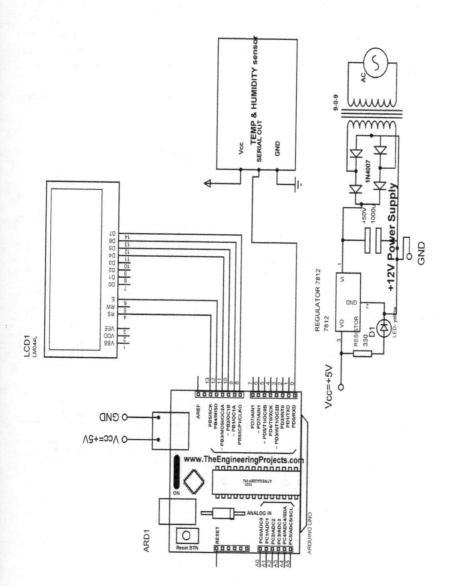

FIGURE 8.11
Proteus simulation model for working of the temperature sensor.

8.6 Humidity/Temperature Sensor—Serial Out

Humidity or temperature sensor from Sunrom with model no. 1211 is used to measure the relative humidity and temperature with the serial interface. The serial data can be checked on PC terminal software at 9600 bps. It can sense temperature from +2°C to +60°C and relative humidity from 1% to 100%. It can be used for environment monitoring at weather stations, greenhouses, and medicine storage.

Figure 8.12 shows the temperature/humidity sensor from (Sunrom model no. 1211).

FIGURE 8.12
Temperature/humidity sensor (Sunrom-1211).

Table 8.3 shows the pin description and Table 8.4 shows the packet format for the sensor model no. 1211.

TABLE 8.3

Pin Description of Humidity/Temperature Sensor-Serial Out (Sunrom-1211)

Pin1	Pin2	Pin3
GND	+5 V	Transmit out-serial data at 9600 bps, 8 bit data, no parity, 1 stop bit.

TABLE 8.4

Packet Format

Byte Count	Details
1	New line character: "\n"
2	Fixed character, "H"
3	Fixed character, ":"
4	Humidity character: "Hundreds"
5	Humidity character: "Tens"
6	Humidity character: "Ones"

(Continued)

TABLE 8.4 (*Continued*)

Packet Format

Byte Count	Details
7	Space
8	Fixed character: "T"
9	Fixed character: ":"
10	Temperature character: "Hundreds"
11	Temperature character: "Tens"
12	Temperature character: "Ones"
13	New line character: "\r"

Example Output of Serial in terminal software: H:058 T:024.

It means relative humidity is 58% and temperature is 24°C.

Figure 8.13 shows the block diagram for the study of temperature/humidity sensor with serial out. It comprises Arduino Uno, LCD (to display the values), power supply, and sensor.

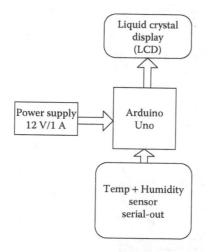

FIGURE 8.13
Block diagram to study temperature/humidity sensor.

8.6.1 Circuit Diagram

8.6.1.1 Temperature Sensor (Serial Out)

To read the temperature sensor serial out, connect the components as follows:

Temperature and humidity sensor connection
- Arduino GND-Module GND
- Arduino +5 V-Module +
- Arduino RX pin-data out pin of sensor

LCD connection

- Arduino digital pin 13-RS pin(4) of LCD
- Arduino digital pin GND-RW pin(5) of LCD
- Arduino digital pin 12-E pin(6) of LCD
- Arduino digital pin 11-D4 pin(11) of LCD
- Arduino digital pin 10-D5 pin(12) of LCD
- Arduino digital pin 9-D6 pin(13) of LCD
- Arduino digital pin 8-D7 pin(14) of LCD

Figure 8.14 shows the circuit diagram for the temperature/humidity sensor interfacing with the Arduino.

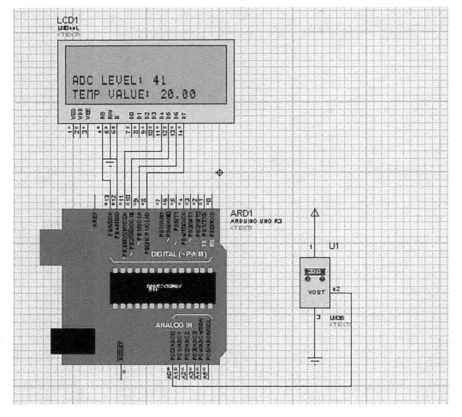

FIGURE 8.14
Circuit diagram to interface the temperature/humidity sensor.

8.6.2 Program

8.6.2.1 Temp/Humidity Sensor—Serial Out

```
#include <LiquidCrystal.h>

LiquidCrystal lcd(13, 12, 11, 10, 9, 8);

String inputString_TEMP_HUMI = "";// a string to hold incoming
data
boolean stringComplete_TEMP_HUMI = false;  // whether the
string is complete

void setup()
{
 Serial.begin(9600);
 lcd.begin(20, 4);
 inputString_TEMP_HUMI.reserve(200); // 200 bytes for the
 inputString
 lcd.setCursor(0,0);
 lcd.print("TEMP_HUMIDITY");
 lcd.setCursor(0,1);
 lcd.print("RADIATION_sensor");
 delay(2000);
}

void loop()
{

 if (stringComplete_TEMP_HUMI)
 {
  lcd.clear();
     Serial.println(inputString_TEMP_HUMI);
  lcd.setCursor(0,2);
  lcd.print("HUM:");
  lcd.print(inputString_TEMP_HUMI[3]);
  lcd.print(inputString_TEMP_HUMI[4]);
  lcd.print(inputString_TEMP_HUMI[5]);
  lcd.setCursor(0,3);
  lcd.print("TEMP:");
  lcd.print(inputString_TEMP_HUMI[9]);
  lcd.print(inputString_TEMP_HUMI[10]);
  lcd.print(inputString_TEMP_HUMI[11]);

  if(inputString_TEMP_HUMI[0]==0x0A)
  {
  lcd.setCursor(0,2);
  lcd.print("HUM:");
  lcd.print(inputString_TEMP_HUMI[4]);
  lcd.print(inputString_TEMP_HUMI[5]);
```

```
lcd.print(inputString_TEMP_HUMI[6]);
lcd.setCursor(0,3);
lcd.print("TEMP:");
lcd.print(inputString_TEMP_HUMI[10]);
lcd.print(inputString_TEMP_HUMI[11]);
lcd.print(inputString_TEMP_HUMI[12]);
}

inputString_TEMP_HUMI = "";
stringComplete_TEMP_HUMI = false;
}
}

void serialEvent()
{
 while (Serial.available())
 {
  char inChar = (char)Serial.read();
  inputString_TEMP_HUMI += inChar;
    if (inChar == 0x0D)
  {
    stringComplete_TEMP_HUMI = true;
  }
 }
}
```

8.7 Light-Dependent Resistor with Liquid Crystal Display

A light-dependent resistor (LDR) or a photo resistor is a device, which measures the intensity of light of the surroundings with resistivity as a function of the incident electromagnetic radiation. Hence, they are light-sensitive devices. They are also known as photoconductive cells, photoconductors, or simply photocells. The resistance of LDR decreases with an increase in the light intensity. It is suitable for using in projects, which require a device or circuit to be automatically switched "on" or "off" in "darkness" or "light."

8.8 Light-Dependent Resistor—Analog Out

The analog output from LDR is connected to the controller to check the light intensity of room and take the corresponding action. Figure 8.15 shows the block diagram of the system and comprises Arduino Uno, LCD, power supply, and LDR.

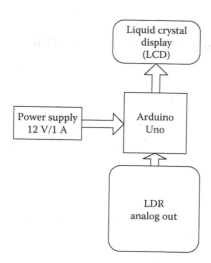

FIGURE 8.15
Block diagram to study the LDR.

Figure 8.16 shows the snapshot for the LDR. To study the behavior of LDR, a small system can be developed. Table 8.5 shows the components list required to implement it.

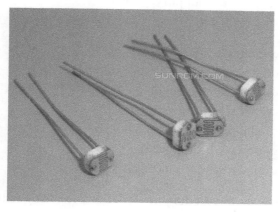

FIGURE 8.16
LDR (Sunrom-3190).

TABLE 8.5

Component List to Study the Working of LDR

Component/Specification	Quantity
Power supply/+12 V/1 A, + 5 V/500 mA	1
Arduino Uno	1
LDR	1
Push button	1
RED LED	1
BLUE LED	1
LCD (20 * 4)	1
LCD patch	1
Sensor patch	1
Connecting wires (M–M, M–F, F–F)	20 each
Zero-size PCB or bread board or a designed PCB	1

8.8.1 Circuit Diagram

To read the LDR analog out, connect the components as follows:

LDR sensor connection
- Arduino GND-Module GND
- Arduino +5 V-Module +
- Arduino A0 pin-data out pin of sensor

LCD connection
- Arduino digital pin 13-RS pin(4) of LCD
- Arduino digital pin GND-RW pin(5) of LCD
- Arduino digital pin 12-E pin(6) of LCD
- Arduino digital pin 11-D4 pin(11) of LCD
- Arduino digital pin 10-D5 pin(12) of LCD
- Arduino digital pin 9-D6 pin(13) of LCD
- Arduino digital pin 8-D7 pin(14) of LCD

Figure 8.17 shows the circuit diagram for LDR interfacing with Arduino.

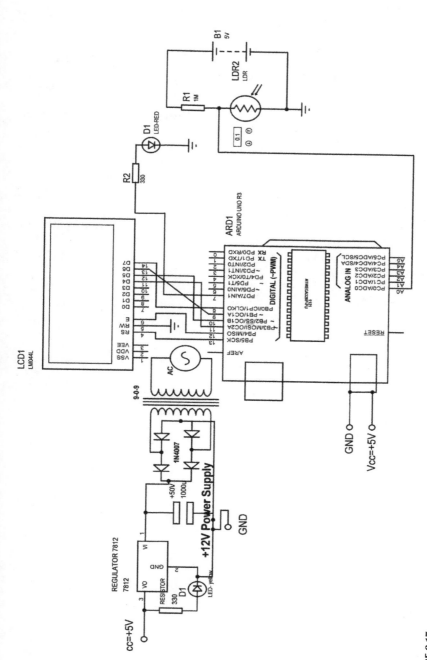

FIGURE 8.17

Circuit diagram for LDR interfacing with the Arduino.

8.8.2 Program

8.8.2.1 Light-Dependent Resistor—Analog Out

```
#include <LiquidCrystal.h>
LiquidCrystal lcd(13, 12, 11, 10,9, 8);// attach LCD pin
RS,E,D4,D5,D6,D7 to the given pins
int LDR_sensor_Pin = A0;     // select the input pin for the
potentiometer
int LDR_sensor_ADC_Value = 0;  // variable to store the value
coming from the sensor
int RED_LED=7;
void setup()
{

  lcd.begin(20, 4);// Initialise 20*4 LCD
  pinMode(RED_LED,OUTPUT);
  lcd.setCursor(0, 0);// set cursor of LCD at column0 and Row0
  lcd.print("LDR based light");
  lcd.setCursor(0, 1);
  lcd.print("intensity monitoring");
  lcd.setCursor(0, 2);
  lcd.print("system at UPES");
  delay(1000);
  lcd.clear();
}

void loop()
{
  LDR_sensor_ADC_Value = analogRead(LDR_sensor_Pin);
  // read the value from the sensor
  lcd.setCursor(0,2);
  lcd.print("ADC LEVEL+LDR:");
  lcd.setCursor(17,2);
  lcd.print(LDR_sensor_ADC_Value);
  if(LDR_sensor_ADC_Value>=100)
  {
  digitalWrite(RED_LED,HIGH);
  delay(20);
  }
  else
  {
  digitalWrite(RED_LED,LOW);
  delay(20);
  }
}
```

8.8.3 Proteus Simulation Model

Figure 8.18 shows the Proteus simulation model for interfacing the LDR with Arduino. Load the program in Arduino and check the working of the circuit. The LCD displays the ADC level of the sensor. Figure 8.18 shows the Proteus simulation model for LDR interfacing with the Arduino.

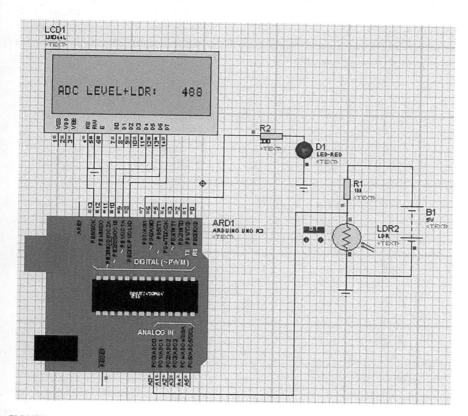

FIGURE 8.18
Proteus simulation model for LDR interfacing.

8.9 Light Intensity Sensor—I2C Out

Digital light sensor BH1750FVI is to be interfaced with the controller in I2C (TWI) mode of communication. This sensor is suitable for obtaining the ambient light intensity value. It is possible to detect a wide range of 0–65535 (Lux) light intensity with high resolution. It operates at 3.3–5 V. Figure 8.19 shows the block diagram to study the light sensor, which comprises of Arduino Uno, LCD, power supply, and the light sensor. Figure 8.20 shows the snapshot of a light sensor.

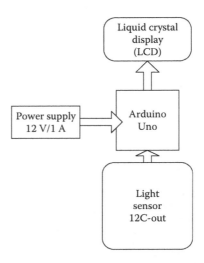

FIGURE 8.19
Block diagram to study the light sensor.

FIGURE 8.20
Light sensor (BH1750FVI).

Table 8.6 shows the component list to study the working of a light sensor.

TABLE 8.6

Component List to Study the Working of a Light Sensor

Component/Specification	Quantity
Power supply/+12 V/1 A, + 5 V/500 mA	1
Arduino Uno	1
Light sensor (BM1750) I2C out	1
LCD (20 * 4)	1
LCD patch	1
Sensor patch	1
Connecting wires (M–M, M–F, F–F)	20 each
Zero-size PCB or bread board or a designed PCB	1

8.9.1 Circuit Diagram

To read the light sensor, connect the components as follows:

Light sensor connection
- Arduino GND-Module GND
- Arduino +5 V-Module +
- Arduino pin5-SCL pin of sensor
- Arduino pin4-SDA pin of sensor

LCD connection
- Arduino digital pin 13-RS pin(4) of LCD
- Arduino digital pin GND-RW pin(5) of LCD
- Arduino digital pin 12-E pin(6) of LCD
- Arduino digital pin 11-D4 pin(11) of LCD
- Arduino digital pin 10-D5 pin(12) of LCD
- Arduino digital pin 9-D6 pin(13) of LCD
- Arduino digital pin 8-D7 pin(14) of LCD

Figure 8.21 shows the circuit diagram of the light sensor interfacing with the Arduino.

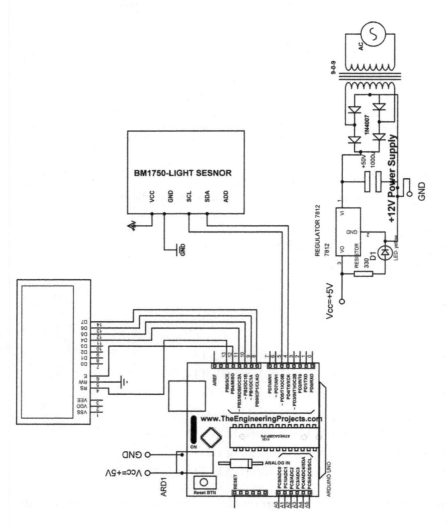

8.9.2 Program

8.9.2.1 LDR TWI (I2C) Out

```
#include <LiquidCrystal.h>
LiquidCrystal lcd(13, 12, 11, 10, 9, 8);
#include <Wire.h>
#include <BH1750.h>

BH1750 lightMeter;

void setup()
{
  lcd.begin(16,4);
  Serial.begin(9600);
  lightMeter.begin();
  lcd.setCursor(0,0);
  lcd.print("LIGHT SENSOR");
  lcd.setCursor(0,1);
  lcd.print("MBM1750 at UPES");

}


void loop()
{
  uint16_t lux = lightMeter.readLightLevel();

  lcd.setCursor(0,2);
  lcd.print("LIGHT LEVEL:");
  lcd.print(lux);
  lcd.print("lux");
  Serial.print("Light level: ");
  Serial.print(lux);
  Serial.println(" lx");
  delay(1000);
}
```

8.10 Servo Motor and the Liquid Crystal Display

A servo motor is a rotary actuator, which allows an angular position control. It comprises a sensor for a position feedback. It has a high torque with metal gears and 360° rotation. It provides 11 kg/cm torque at 4.8 V, 13.5 kg/cm at 6 V, and 16 kg/cm at 7.2 V. Figure 8.22 shows the block diagram to study the behavior of a servo motor. Figure 8.23 shows the snapshot for the servo motor from Robokits. Table 8.7 shows the component list to study the working of a servo motor.

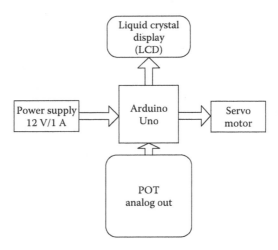

FIGURE 8.22
Block diagram to study a servo motor.

FIGURE 8.23
Servo motor from Robokits.

TABLE 8.7

Component List to Study the Working of a Servo Motor

Component/Specification	Quantity
Power supply/+12 V/1 A, + 5 V/500 mA	1
Arduino Uno	1
Servo motor	1
LCD (20 * 4)	1
POT	1
LCD patch	1
Sensor patch	1
Connecting wires (M–M, M–F, F–F)	20 each
Zero-size PCB or bread board or a designed PCB	1

8.10.1 Circuit Diagram

To read the servo motor, connect the components as follows:

Servo connection
- Arduino GND-Module GND
- Arduino +5 V-Module +
- Arduino pin(3)-servo PWM pin

POT connection
- Arduino GND-Module GND
- Arduino +5 V-Module +
- Arduino A0 pin-data out pin of POT

LCD connection
- Arduino digital pin 13-RS pin(4) of LCD
- Arduino digital pin GND-RW pin(5) of LCD
- Arduino digital pin 12-E pin(6) of LCD
- Arduino digital pin 11-D4 pin(11) of LCD
- Arduino digital pin 10-D5 pin(12) of LCD
- Arduino digital pin 9-D6 pin(13) of LCD
- Arduino digital pin 8-D7 pin(14) of LCD

Figure 8.24 shows the circuit diagram to interface servo motor with the Arduino.

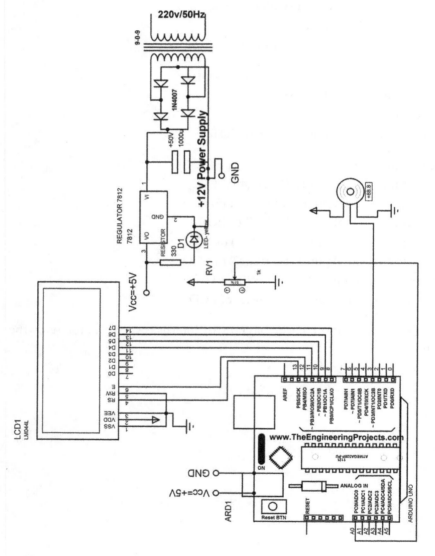

FIGURE 8.24
Circuit diagram to interface servo motor.

8.10.2 Program

```
#include <Servo.h>
#include <LiquidCrystal.h>
LiquidCrystal lcd(13, 12, 11, 10, 9, 8);
Servo myservo;
 int POT_PIN = A0;  // analog pin used to connect the
 potentiometer
 int POT_PIN_ADC_LEVEL;    // variable to read the value from
 the analog pin

void setup()
{   myservo.attach(3);  // connect the servo on pin 9 to the
servo object
 lcd.begin(20,4);
 lcd.setCursor(0, 0);
 lcd.print("Servo ANALOG write ");
 lcd.setCursor(0, 1);
 lcd.print("system at UPES....");
}

void loop()
{
 POT_PIN_ADC_LEVEL = analogRead(POT_PIN);                // reads
 the value of the potentiometer (value between 0 and 1023)

 POT_PIN_ADC_LEVEL = map(POT_PIN_ADC_LEVEL, 0, 1023, 0, 179);
 // scale it to use it with the servo (value between 0 and 180)

 myservo.write(POT_PIN_ADC_LEVEL);   // sets the servo
 position according to the scaled value
 lcd.setCursor(0, 2);
 lcd.print("ANGLE:");
 lcd.print(POT_PIN_ADC_LEVEL);
 delay(15);
}
```

8.10.3 Proteus Simulation Model

Figure 8.25 shows the Proteus simulation model for interfacing the servo motor with the Arduino. Load the program in the Arduino and check the working of the circuit. The LCD displays the angle of servo.

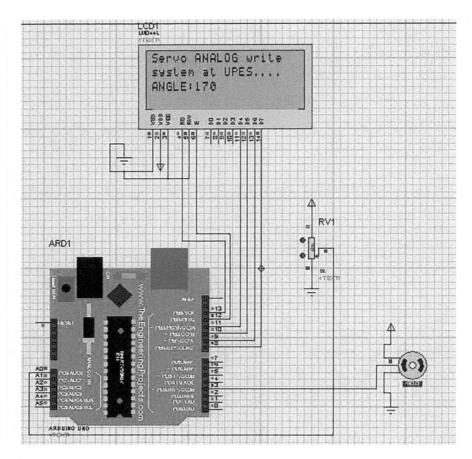

FIGURE 8.25
Proteus simulation model for servo motor interfacing with Arduino.

9

Arduino and Motors/Actuators

This chapter describes the Arduino interfacing with the motors or actuators such as DC motor, stepper motor, and the AC motor. The working of the devices is discussed with the help of an interfacing circuit, program, and Proteus simulation models.

9.1 DC Motor

DC motor with 100 rpm is used to demonstrate the working. It is very easy to use and is available in standard size. It has 1.2 kg/cm torque, No-load current = 60 mA (Max), and Load current = 300 mA (Max). Figure 9.1 shows the DC motor from Robokits India.

Figure 9.2 shows the block diagram to study the working of a DC motor. Block diagram comprises the Arduino Uno, liquid crystal display (LCD), power supply, DC motor, and motor driver. As DC motor is 12 V/1 A, so L293D motor driver is used.

FIGURE 9.1
DC motor from Robokits India.

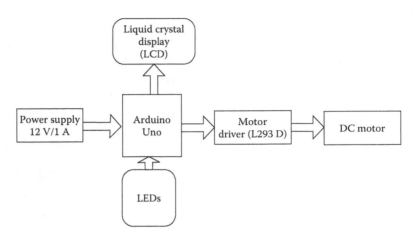

FIGURE 9.2
Block diagram to study a DC motor.

Table 9.1 shows the component list for interfacing the DC motor with the Arduino.

TABLE 9.1

Component List to Study the Working of a DC Motor

Component/Specification	Quantity
Power supply/+12 V/1 A, +5 V/500 mA	1
Arduino Uno	1
DC motor	1
LCD (20 * 4)	1
LCD patch	1
L293D	1
L293D patch	1
LED (different color)	4
Connecting wires (M–M, M–F, F–F)	20 each
Zero size PCB or bread board or designed PCB	1

9.1.1 Circuit Diagram

To control the DC motor, connect the components as follows:

L293D and DC motor connection
- L293D pin 3-to +ve pin of DC motor1
- L293D pin 6-to −ve pin of DC motor1
- L293D pin 11-to +ve pin of DC motor2
- L293D pin 14-to +ve pin of DC motor2

L293D connection
- Arduino GND-4,5,12,13 pins of IC
- Arduino +5 V-1,9,16 pins of IC
- Arduino pin 7-pin 2 of IC
- Arduino pin 6-pin 7 of IC
- Arduino pin 5-pin 10 of IC
- Arduino pin 4-pin 15 of IC
- L293D pin8-+ve of 12 V battery

LED connection
- Arduino pin 7-LED1
- Arduino pin 6-LED2
- Arduino pin 5-LED3
- Arduino pin 4-LED4

LCD connection
- Arduino digital pin 13-RS pin(4) of LCD
- Arduino digital pin GND-RW pin(5) of LCD
- Arduino digital pin 12-E pin(6) of LCD
- Arduino digital pin 11-D4 pin(11) of LCD
- Arduino digital pin 10-D5 pin(12) of LCD
- Arduino digital pin 9-D6 pin(13) of LCD
- Arduino digital pin 8-D7 pin(14) of LCD

Figure 9.3 shows the circuit diagram to interface the DC motor with the Arduino.

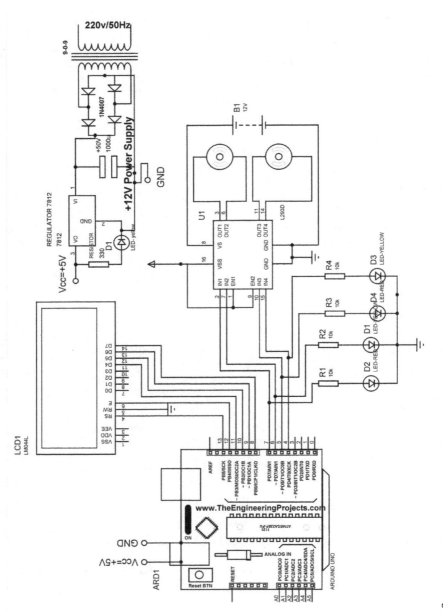

FIGURE 9.3
Circuit diagram to interface DC motor.

9.1.2 Program

```
#include <LiquidCrystal.h>
LiquidCrystal lcd(13, 12, 11, 10, 9, 8);
int MPIN1= 7;
int MPIN2= 6;
int MPIN3= 5;
int MPIN4= 4;

void setup()
{
 pinMode(MPIN1, OUTPUT);
 pinMode(MPIN2, OUTPUT);
 pinMode(MPIN3, OUTPUT);
 pinMode(MPIN4, OUTPUT);
 lcd.begin(20,4);
 lcd.setCursor(0, 0);
 lcd.print("DC Motor direction");
 lcd.setCursor(0, 1);
 lcd.print("control system...");
 delay(1000);
 lcd.clear();
}

// the loop routine runs over and over again forever:
void loop()
{
 digitalWrite(MPIN1, HIGH);
 digitalWrite(MPIN2, LOW);
 digitalWrite(MPIN3, HIGH);
 digitalWrite(MPIN4, LOW);
 lcd.setCursor(0, 2);
 lcd.print("CLOCKWISE");
 delay(2000);
 lcd.clear();// wait for a second
 digitalWrite(MPIN1, LOW);
 digitalWrite(MPIN2, HIGH);
 digitalWrite(MPIN3, LOW);
 digitalWrite(MPIN4, HIGH);
 lcd.setCursor(0, 2);
 lcd.print("ANTI-CLOCKWISE");
 delay(2000);             // wait for a second
 lcd.clear();
 digitalWrite(MPIN1, LOW);
 digitalWrite(MPIN2, LOW);
 digitalWrite(MPIN3, HIGH);
 digitalWrite(MPIN4, LOW);
 lcd.setCursor(0, 2);
 lcd.print("LEFT");
```

```
delay(2000);
lcd.clear();
digitalWrite(MPIN1, HIGH);
digitalWrite(MPIN2, LOW);
digitalWrite(MPIN3, LOW);
digitalWrite(MPIN4, LOW);
lcd.setCursor(0, 2);
lcd.print("RIGHT");
delay(2000);
lcd.clear();
}
```

9.1.3 Proteus Simulation Model

Figures 9.4 and 9.5 shows the Proteus simulation model for interfacing the DC motor with the Arduino. Load the program in the Arduino and check the working of the circuit. The LCD displays the direction of the DC motor.

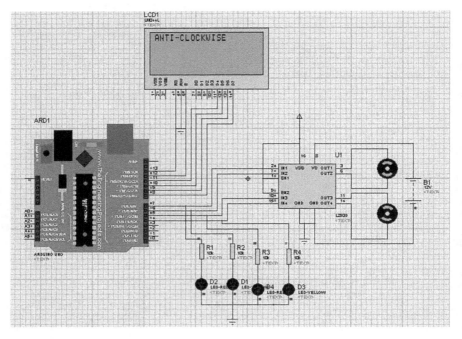

FIGURE 9.4
Proteus simulation model to control DC motor (anticlockwise).

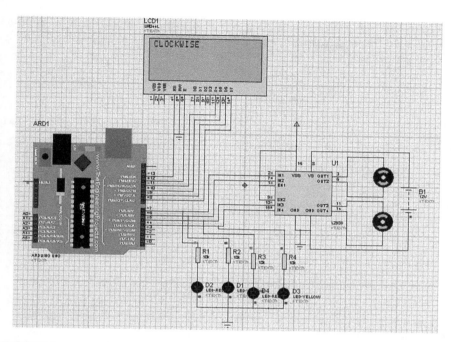

FIGURE 9.5
Proteus simulation model to control a DC motor (clockwise).

9.2 Stepper Motor

To demonstrate the working of stepper motor, stepper motor from Robokits INDIA is considered. A four-wire bipolar stepper motor can take a step angle of 3.75° and operates on 3.3 V. It has a holding torque of 0.65 Kg/cm at 0.6A per winding with a weight of 180 g. Figure 9.6 shows the snapshot of a stepper motor from Robokits India.

FIGURE 9.6
Stepper motor from Robokits India.

Figure 9.7 shows the block diagram to study the working of a stepper motor. It comprises Arduino Uno, power supply, LCD, stepper motor, and a motor driver (L293D).

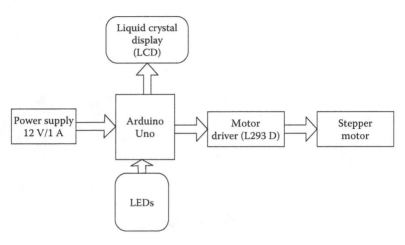

FIGURE 9.7
Block diagram to study the working of a stepper motor.

Table 9.2 shows the component list for interfacing stepper motor with the Arduino.

TABLE 9.2

Component List to Study the Working of a Stepper Motor

Component/Specification	Quantity
Power supply/+12 V/1 A, +5 V/500 mA	1
Arduino Uno	1
Stepper motor	1
LCD (20 * 4)	1
LCD patch	1
L293D	1
L293D patch	1
LED (different color)	4
Connecting wires (M–M, M–F, F–F)	20 each
Zero-size PCB or bread board or a designed PCB	1

9.2.1 Circuit Diagram

To control the stepper motor, connect the components as follows:

L293D and stepper motor connection
- L293D pin 3-to +ve pin of DC motor1
- L293D pin 6-to −ve pin of DC motor1
- L293D pin 11-to +ve pin of DC motor2
- L293D pin 14-to +ve pin of DC motor2

L293D connection
- Arduino GND-4,5,12,13 pins of IC
- Arduino +5 V-1,9,16 pins of IC
- Arduino pin 7-pin 2 of IC
- Arduino pin 6-pin 7 of IC
- Arduino pin 5-pin 10 of IC
- Arduino pin 4-pin 15 of IC
- L293D pin8-+ve of 12 V battery

LED connection
- Arduino pin 7-LED1
- Arduino pin 6-LED2
- Arduino pin 5-LED3
- Arduino pin 4-LED4

LCD connection
- Arduino digital pin 13-RS pin(4) of LCD
- Arduino digital pin GND-RW pin(5) of LCD
- Arduino digital pin 12-E pin(6) of LCD
- Arduino digital pin 11-D4 pin(11) of LCD
- Arduino digital pin 10-D5 pin(12) of LCD
- Arduino digital pin 9-D6 pin(13) of LCD
- Arduino digital pin 8-D7 pin(14) of LCD

Figure 9.8 shows the circuit diagram to interface the stepper motor with the Arduino.

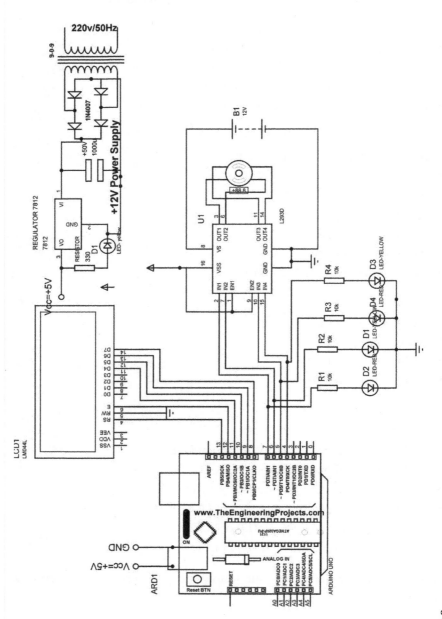

FIGURE 9.8
Circuit diagram to interface stepper motor.

9.2.2 Program

```
#include <LiquidCrystal.h>

// initialize the library with the numbers of the
interface pins
LiquidCrystal lcd(13, 12, 11, 10, 9, 8);

int A_Phase_MPIN1= 7;
int B_Phase_MPIN2= 6;
int C_Phase_MPIN3= 5;
int D_Phase_MPIN4= 4;

void setup()
{
  pinMode(A_Phase_MPIN1, OUTPUT);
  pinMode(B_Phase_MPIN2, OUTPUT);
  pinMode(C_Phase_MPIN3, OUTPUT);
  pinMode(D_Phase_MPIN4, OUTPUT);
  lcd.begin(20,4);
  lcd.setCursor(0, 0);
  lcd.print("DC Motor direction");
  lcd.setCursor(0, 1);
  lcd.print("control system...");
  delay(2000);
  lcd.clear();
}

// the loop routine runs over and over again forever:
void loop()
{
  digitalWrite(A_Phase_MPIN1, HIGH);
  digitalWrite(B_Phase_MPIN2, LOW);
  digitalWrite(C_Phase_MPIN3, LOW);
  digitalWrite(D_Phase_MPIN4, LOW);
  lcd.setCursor(0, 2);
  lcd.print("90 Degree");
  delay(2000);
  lcd.clear();// wait for a second
  digitalWrite(A_Phase_MPIN1, LOW);
  digitalWrite(B_Phase_MPIN2, HIGH);
  digitalWrite(C_Phase_MPIN3, LOW);
  digitalWrite(D_Phase_MPIN4, LOW);
  lcd.setCursor(0, 2);
  lcd.print("180 Degree");
  delay(2000);              // wait for a second
  lcd.clear();
  digitalWrite(A_Phase_MPIN1, LOW);
  digitalWrite(B_Phase_MPIN2, LOW);
  digitalWrite(C_Phase_MPIN3, HIGH);
```

```
digitalWrite(D_Phase_MPIN4, LOW);
lcd.setCursor(0, 2);
lcd.print("270 Degree");
delay(2000);
lcd.clear();
digitalWrite(A_Phase_MPIN1, LOW);
digitalWrite(B_Phase_MPIN2, LOW);
digitalWrite(C_Phase_MPIN3, LOW);
digitalWrite(D_Phase_MPIN4, HIGH);
lcd.setCursor(0, 2);
lcd.print("360 Degree");
delay(2000);
lcd.clear();
}
```

9.2.3 Proteus Simulation Model

Figure 9.9 shows the Proteus simulation model for interfacing the stepper motor with the Arduino. Load the program in Arduino and check the working of the circuit. The LCD displays the angle of a stepper motor.

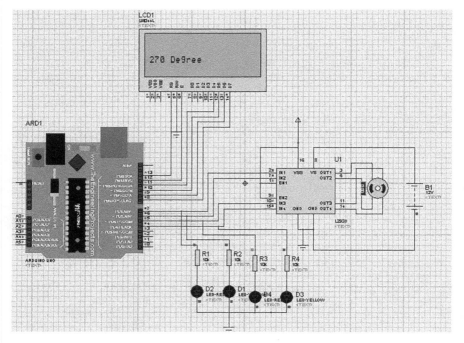

FIGURE 9.9
Proteus simulation model to control a stepper motor.

9.3 AC Motor with Relay

An AC motor is an induction motor in which the electric current in the rotor is required to produce torque. Torque is obtained by an electromagnetic induction from the magnetic field of the stator winding. Figure 9.10 shows the block diagram to study the AC motor. It comprises the Arduino, AC motor, LCD, and the power supply.

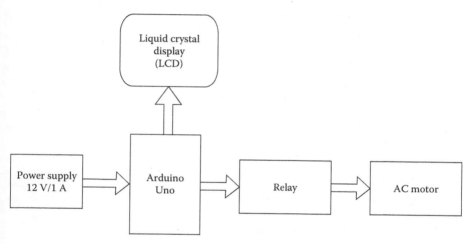

FIGURE 9.10
Block diagram to study an AC motor.

Table 9.3 shows the component list for interfacing AC motor with the Arduino.

TABLE 9.3

Component List to Study the Working of a AC Motor

Component/Specification	Quantity
Power supply/+12 V/1 A, + 5 V/500 mA	1
Arduino Uno	1
AC motor	1
LCD (20 * 4)	1
LCD patch	1
Relay	1
Relay patch	1
2N2222	1
LED (different color)	4
Connecting wires (M–M, M–F, F–F)	20 each
Zero-size PCB or bread board or a designed PCB	1

9.3.1 Circuit Diagram

To control the AC motor, connect the components as follows:

2N2222, Relay, and Arduino connection
- Arduino pin 7-base of 2N2222
- Collector of 2N2222-L2 end relay
- L1 end of relay to +12 V of battery or power supply
- COM pin of relay-one end of AC motor
- NO pin of relay-other end of AC motor

LCD connection
- Arduino digital pin 13-RS pin(4) of LCD
- Arduino digital pin GND-RW pin(5) of LCD
- Arduino digital pin 12-E pin(6) of LCD
- Arduino digital pin 11-D4 pin(11) of LCD
- Arduino digital pin 10-D5 pin(12) of LCD
- Arduino digital pin 9-D6 pin(13) of LCD
- Arduino digital pin 8-D7 pin(14) of LCD

Figure 9.11 shows the circuit diagram to interface the AC motor with the Arduino through the relay and the transistor.

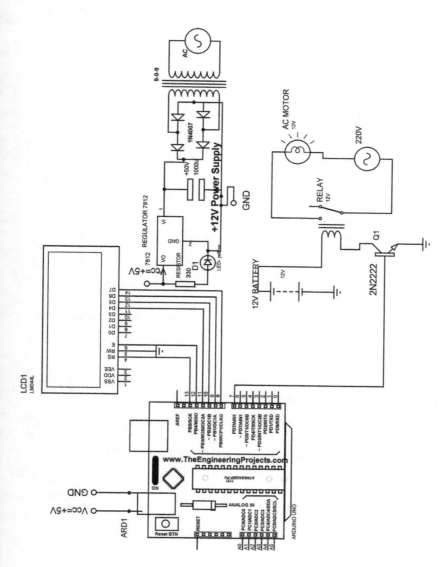

FIGURE 9.11
Circuit diagram to interface AC motor.

9.3.2 Program

```
#include <LiquidCrystal.h>
LiquidCrystal lcd(13, 12, 11, 10, 9, 8);
int RELAY_2N2222= 7;

void setup()
{
  lcd.begin(20,4);
  pinMode(RELAY_2N2222, OUTPUT);
  lcd.setCursor(0, 0);
  lcd.print("AC MOTOR ON/OFF");
  lcd.setCursor(0, 1);
  lcd.print("System at UPES");
  delay(1000);
}

void loop()
{
 lcd.setCursor(0, 2);
 lcd.print("AC MOTOR ONN");
 digitalWrite(RELAY_2N2222, HIGH);   // turn the LED on
 (HIGH is the voltage level)
 delay(2000);         // wait for a second

 lcd.setCursor(0, 2);
 lcd.print("AC MOTOR OFF");
 digitalWrite(RELAY_2N2222, LOW);    // turn the LED off by
 making the voltage LOW
 delay(2000);         // wait for a second
}
```

9.3.3 Proteus Simulation Model

Figures 9.12 and 9.13 show the Proteus simulation model for interfacing the AC motor with the Arduino. Load the program in the Arduino and check the working of the circuit. The LCD displays the status of a stepper motor.

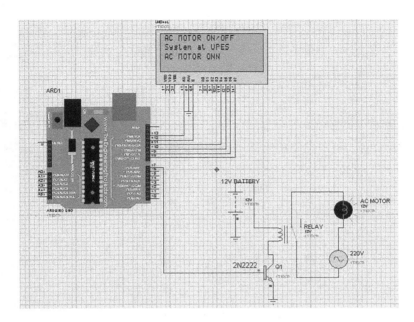

FIGURE 9.12
Proteus simulation model to control AC motor (ON condition).

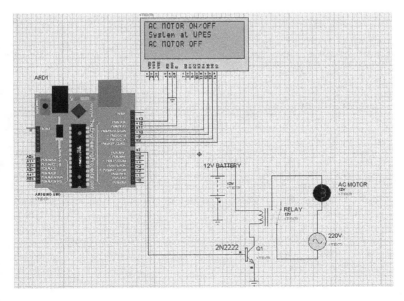

FIGURE 9.13
Proteus simulation model to control AC motor (OFF condition).

Section III

Arduino and Wireless Communication

10

Arduino and Wireless Communication

This chapter describes the Arduino interfacing with the wireless communication modules such as 2.4 GHz RF modem and global system for mobile (GSM). The working of the devices is discussed with the help of an interfacing circuit, program, and Proteus simulation models.

10.1 RF Modem (2.4 GHz)

2.4 GHz RF modem from Sunrom1418 operates at 3.3 or 5 V. RF modem can be used for a two-way wireless communication. It has a bidirectional UART serial data transfer with a half-duplex capability. It operates on 2.4 GHz, which is a license-free band. It can be used for applications such as sensor networks, smart houses, and so on. It has an adjustable baud rate of 9600/4800/2400/19200 bps. To set the baud rate the unit should be in "OFF" condition. To adjust baud rate the two jumpers called B1 and B2 are given onboard, which are generally open to be at a baud rate of 9600 bps. To make any jumper on, short that jumper by soldering its pads. Baud rate can be configured as follows:

B1 = OFF, B2 = OFF, 9600 bps (Default)
B1 = ON, B2 = OFF, 4800 bps
B1 = OFF, B2 = ON, 2400 bps
B1 = ON, B2 = ON, 19200 bps

Frequency channel setting can be used to have multiple sets operating at the same time but without interfering with each other. The pair with the same channel setting can communicate with each other. Frequency channel has to be set when the unit is OFF. Frequency channel can be configured as follows:

F1 = OFF, F2 = OFF, Channel #1 (Default)
F1 = ON, F2 = OFF, Channel #2
F1 = OFF, F2 = ON, Channel #3
F1 = ON, F2 = ON, Channel #4

Figure 10.1 shows the snapshot of the 2.4 GHz RF modem (Table 10.1).

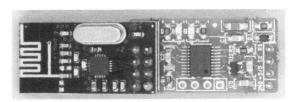

FIGURE 10.1
2.4 GHz RF modem (Sunrom model-1418).

TABLE 10.1

Pin Description of RF Modem

RXD	Receive input (Connected to TXD pin of the microcontroller)
TXD	Transmit output (Connected to RXD pin of the microcontroller)
+3/5 V	Regulated 3.3–5 V supply input
GND	Ground (Must be connected with the ground of microcontroller)

To study the working of RF modem a basic system can be designed to check the feasibility of the modem. Figure 10.2 shows the block diagram to design the system with two sections, that is, the transmitter and the receiver with a wireless link between them.

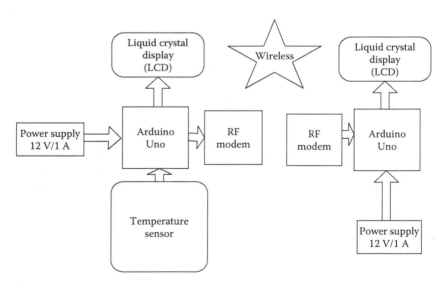

FIGURE 10.2
Block diagram to study the RF modem.

Transmitter section: It comprises Arduino, power supply, liquid crystal display (LCD) (to display readings from the sensor), temperature sensor (to sense the surrounding temperature), and RF modem to transmit the data to the receiver.

Receiver section: It comprises Arduino, LCD, power supply, and RF modem to receive data from the transmitter.

Table 10.2 shows the components that are required to study the operation of a RF modem.

TABLE 10.2

Components List to Study the Working of a RF Modem

Component/Specification	Quantity
Power supply/+12 V/1 A	2
Arduino Uno	2
Temperature sensor	1
RF modem (2.4 GHz)	1 pair
RF modem patch	2
LCD (20 * 4)	2
LCD patch	2
Connecting wires (M–M, M–F, F–F)	30 each
Zero-size PCB or bread board or a designed PCB	2

10.1.1 Circuit Diagram

10.1.1.1 Transmitter Section

Arduino and LM35

- Arduino GND-Module GND
- Arduino +5 V-Module +
- Arduino A0 pin-data out pin of sensor

Arduino and RF modem

- Arduino pin0(TX)-RX pin of RF modem
- Arduino pin0(RX)-Tx pin of RF modem
- Arduino+5 V-+Vcc of RF modem
- Arduino Ground-GND of RF modem

Arduino and LCD

- Arduino digital pin 13-RS pin(4) of LCD
- Arduino digital pin GND-RW pin(5) of LCD

- Arduino digital pin 12-E pin(6) of LCD
- Arduino digital pin 11-D4 pin(11) of LCD
- Arduino digital pin 10-D5 pin(12) of LCD
- Arduino digital pin 9-D6 pin(13) of LCD
- Arduino digital pin 8-D7 pin(14) of LCD

10.1.1.2 Receiver Section

Arduino and RF modem

- Arduino pin0(TX)-RX pin of RF modem
- Arduino pin0(RX)-Tx pin of RF modem
- Arduino+5 V-+Vcc of RF modem
- Arduino Ground-GND of RF modem

Arduino and LCD

- Arduino digital pin 13-RS pin(4) of LCD
- Arduino digital pin GND-RW pin(5) of LCD
- Arduino digital pin 12-E pin(6) of LCD
- Arduino digital pin 11-D4 pin(11) of LCD
- Arduino digital pin 10-D5 pin(12) of LCD
- Arduino digital pin 9-D6 pin(13) of LCD
- Arduino digital pin 8-D7 pin(14) of LCD

Figure 10.3 shows the circuit diagram for the transmitter and the receiver sections.

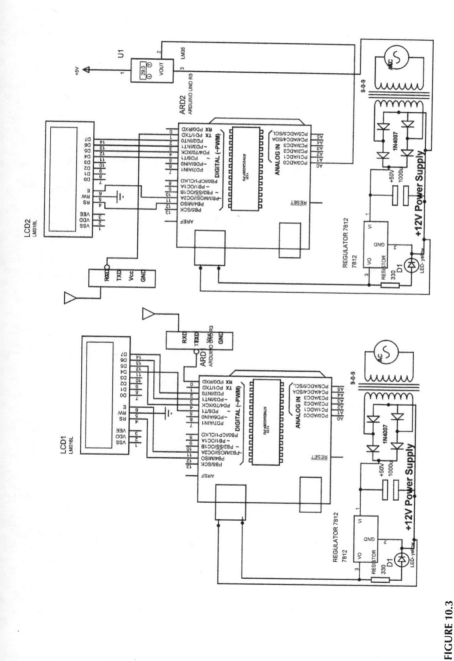

FIGURE 10.3
Circuit diagram for transmitter and receiver sections.

10.1.2 Program

10.1.2.1 Transmitter Program

```
#include <LiquidCrystal.h>
LiquidCrystal lcd(13, 12, 11, 10, 9, 8);
int TEMP_SENSOR=A0;
void setup()
{
 lcd.begin(20,4);
 Serial.begin(9600);// initialize serial communication at
 9600 bits per second
 lcd.setCursor(0,0);
 lcd.print("wireless Serial ");
 lcd.setCursor(0,1);
 lcd.print("communication via 2.4 GHz");
 lcd.setCursor(0,2);
 lcd.print("modem at SME.....");
}

void loop()
{
 int TEMP_SENSOR_ADC_VALUE = analogRead(TEMP_SENSOR);
 int TEMP_VAL=TEMP_SENSOR_ADC_VALUE/2;//10mV/0C is resolution
 of sensor
 lcd.setCursor(0,3);
 lcd.print("Temp. in 0C:");
 lcd.setCursor(13,3);
 lcd.print(TEMP_VAL);// print temperature value on LCD
 Serial.write(TEMP_VAL);
 delay(100);
}
```

10.1.2.2 Receiver Program

```
#include <LiquidCrystal.h>
LiquidCrystal lcd(13, 12, 11, 10, 9, 8);

void setup()
{
 lcd.begin(20,4);
 Serial.begin(9600);// initialize serial communication at 9600
 bits per second
 lcd.setCursor(0,0);
 lcd.print("wireless Serial ");
 lcd.setCursor(0,1);
 lcd.print("communication via 2.4 GHz");
 lcd.setCursor(0,2);
 lcd.print("modem at SME.....");
}
```

```
// the loop routine runs over and over again forever:
void loop()
{
  int TEMP_VALUE=Serial.read();
  lcd.setCursor(0,3);
  lcd.print("Temp. in 0C:");
  lcd.setCursor(13,3);
  lcd.print(TEMP_VALUE);// print temperature value on LCD
  delay(100);

  Serial.print("TEMP_VAL:");
  Serial.println(TEMP_VALUE);
  delay(100);     // delay in between reads for stability
}
```

10.1.3 Proteus Simulation Model

Connect the components in virtual environment as in the circuit diagram. Figures 10.4 and 10.5 shows the Proteus simulation model for the transmitter and the receiver sections. Load the program in Arduino and check the working of the circuit. The serial data can be checked on the virtual terminal. For this, connect the RX pin of the virtual terminal to the TX pin of the Arduino as shown in Figure 10.5.

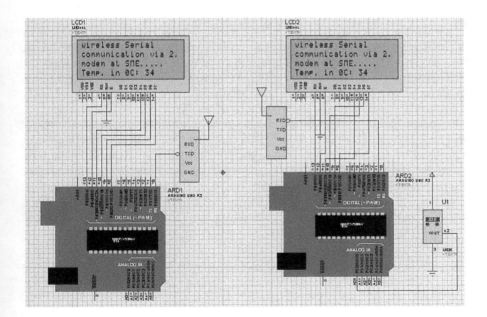

FIGURE 10.4
Proteus simulation model for the transmitter and receiver.

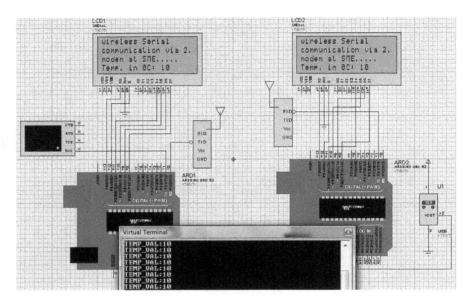

FIGURE 10.5
Proteus simulation model with virtual terminal.

10.2 Global System for Mobile Modem

A GSM is to communicate with the mobile phone. It can be connected in two modes with controller—TTL (to interface with the microcontrollers) and RS232 (to interface directly to the PC). It is preferable to take TTL logic GSM modem. GSM need 12 V/1 A DC power supply.

Figure 10.6 shows the snapshot of a GSM modem.

Figure 10.7 shows the block diagram to study the working of a GSM modem. It comprises two parts: (1) transmitter and the (2) mobile phone. Transmitter section comprises Arduino Uno, power supply, LCD, temperature sensor, and GSM mode. The system is designed to capture the surrounding temperature reading and send it to a predefined mobile number.

FIGURE 10.6
GSM modem.

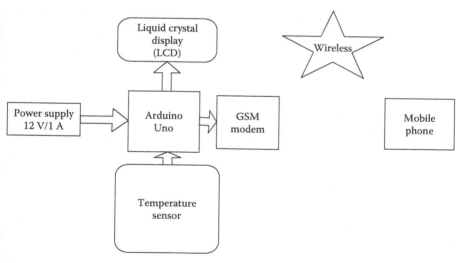

FIGURE 10.7
Block diagram to study the GSM modem.

Table 10.3 describes the component list of a GSM system.

TABLE 10.3

Components List to Study the Working of a GSM Modem

Component/Specification	Quantity
Power supply/+12 V/1 A	1
Arduino Uno	1
Push button	1
GSM modem	1
GSM modem patch	1
LCD (20 * 4)	1
LCD patch	1
Connecting wires (M–M, M–F, F–F)	20 each
Zero-size PCB or bread board or a designed PCB	1

10.2.1 Circuit Diagram

To read the GSM modem with the Arduino, connect the components as follows:
A push button is used to generate signal, which when pressed a message will be delivered.

Arduino and push button

- Arduino pin7-connect message button
- Arduino pin6-connect CALL button
- Arduino pin5-connect END button

Arduino and GSM modem

- Arduino pin0(RX)-Tx pin of modem
- Arduino pin0(TX)-RX pin of modem
- Arduino+5 V-+Vcc of GSM modem
- Arduino Ground-GND of GSM modem

Arduino and LCD

- Arduino digital pin 13-RS pin(4) of LCD
- Arduino digital pin GND-RW pin(5) of LCD
- Arduino digital pin 12-E pin(6) of LCD
- Arduino digital pin 11-D4 pin(11) of LCD
- Arduino digital pin 10-D5 pin(12) of LCD
- Arduino digital pin 9-D6 pin(13) of LCD
- Arduino digital pin 8-D7 pin(14) of LCD

Figure 10.8 shows the circuit diagram for the system.

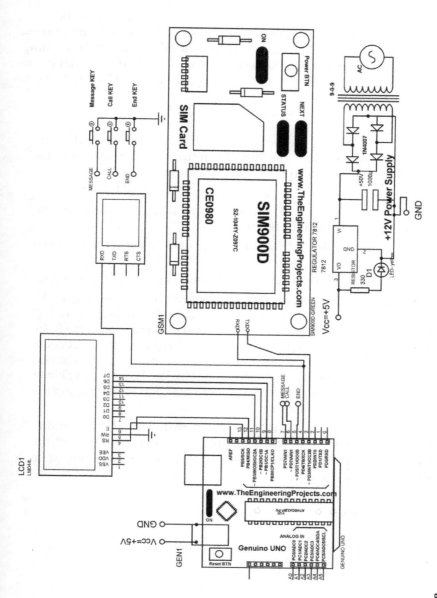

FIGURE 10.8
Circuit diagram.

10.2.2 Program

To program GSM modem, make pin9 and pin10 as transmitter and receiver and include *SoftwareSerial* library. The format is *SoftwareSerial mySerial (Rx, Tx)*. In this code, pin9 of Arduino is the Rx and pin10 of Arduino is the Tx.

The first step is to *set baud rate* of SoftwareSerial library to communicate with the GSM module. This is achieved by the command *mySerial.begin* function. The second step is to set the baud rate of Arduino IDE's serial monitor. This is done by the command *Serial.begin* function. The baud rate is 9600 bits/s is set.

The details of some of the important instructions are as follows:

Serial.available(): It checks for any data that are received to the serial port of Arduino. The function returns the number of bytes that are available to read from the serial buffer.

Serial.read(): It reads the data available at the serial buffer.

mySerial.available(): It checks for the data received from the GSM module through the SoftwareSerial.

mySerial.read(): It reads the incoming data at the software serial port.

Serial.write(): It prints the data to the serial monitor of the Arduino.

SendMessage(): It sends an SMS. This is achieved by sending an AT Command "AT+CMGF=1." To achieve this, the mySerial.println() function is used. The mobile number is to be added to send the SMS. This is achieved with an AT command "AT+CMGS=\"+91xxxxxxxx xx\"\r"—where x is to be replaced with the mobile number. The end of the SMS content is identified with CTRL+Z. The ASCII value of this CTRL+Z is 26. So, char(26) is sent to the GSM module using the line mySerial.println((char)26).

RecieveMessage(): It is used to receive a SMS. The AT command to receive a SMS is "AT+CNMI=2,2,0,0,0." AT Command to read the SMS stored in SIM card is "AT+CMGL=\"ALL\"\r" which is to be sent to the GSM module.

10.2.2.1 Main Program

```
#include <LiquidCrystal.h>
LiquidCrystal lcd(13, 12, 11, 10, 9, 8);

#include <SoftwareSerial.h>
SoftwareSerial mySerial(3, 4);
#define MESSAGE_button 7
#define CALL_button 6
#define END_button 5
String number ="9837043685"; // Add the 10-Digit Mobile Number
```

```
void setup()
{
  lcd.begin(20, 4);
  Serial.begin(9600);
  mySerial.begin(9600);
  pinMode(MESSAGE_button,INPUT_PULLUP);
  pinMode(CALL_button,INPUT_PULLUP);
  pinMode(END_button,INPUT_PULLUP);
  //digitalWrite(msg_key,HIGH);
  //digitalWrite(call_key,HIGH);
  //digitalWrite(end_key,HIGH);
  lcd.setCursor(0, 0);
  lcd.print("GSM MESSAGE ");
  lcd.setCursor(0, 1);
  lcd.print("sending system.... ");
  delay(1000);
}
void loop()
{
  //Sends an sms everytime msg_key is pressed
  if (digitalRead(MESSAGE_button)==LOW) // Check if the
  sms key is being pressed
  {
    lcd.clear();
    mySerial.println("AT+CMGF=1"); // Set the Mode as Text Mode
    lcd.setCursor(0, 2);
    lcd.print("AT+CMGF=1");
    delay(150);
    mySerial.println("AT+CMGS=\"+919837043685\"\r");
    // Specify the Destination number in
    international format by replacing the 0's
    lcd.setCursor(0, 2);
    lcd.print("AT+CMGS=\"+919837043685\"\r");
    delay(150);
    mySerial.print("message Send"); // Enter the message
    lcd.setCursor(0, 2);
    lcd.print("message Send");
    delay(150);
    mySerial.write((byte)0x1A); // End of message character
    0x1A : Equivalent to Ctrl+z
    lcd.setCursor(0, 2);
    lcd.print("cntrl+Z");
    delay(50);
    mySerial.println();
  }
  //Makes a call when call_key is pressed
  else if (digitalRead(CALL_button)==LOW) // Check if the call
  key is being pressed
  {
    lcd.clear();
```

```
  lcd.setCursor(0, 2);
  lcd.print("ATD+91"+number+";");
  mySerial.println("ATD+91"+number+";"); //Specify the
  number to call
  while(digitalRead(CALL_button)==LOW);
  delay(50);
}
//Hang the call
else if (digitalRead(END_button)==LOW) // Check if the hang
up key is being pressed
{
  lcd.clear();
  lcd.setCursor(0, 2);
  lcd.print("ATH");
  mySerial.println("ATH");
  while(digitalRead(END_button)==LOW);
  delay(50);
}
}
```

10.2.3 Proteus Simulation Model

Connect the components in the virtual environment as in the circuit diagram. Figure 10.9 shows the Proteus simulation model for the transmitter section. Load the program in the Arduino and check the working of the circuit. The serial data can be checked on the virtual terminal. For this, connect RX pin of virtual terminal to TX pin of Arduino as shown in Figure 10.9.

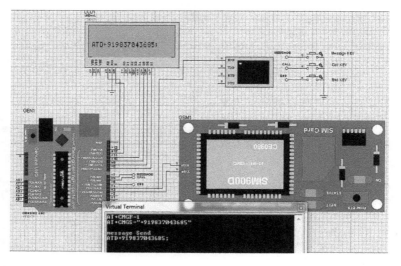

FIGURE 10.9
Proteus simulation model.

Section IV

Projects

11

2.4 GHz RF Modem-Based Security System for Restricted Area

11.1 Introduction

This project is designed for monitoring the security of the restricted areas. The system is designed in two sections: (1) sensor node and (2) server. It is designed in such a way that it will sense any intrusion with the help of a passive infrared (PIR) sensor at the sensor node and communicated to the server wirelessly with a 2.4 GHz RF modem. At the server, a LabVIEW GUI is created to observe the system. Figure 11.1 shows the block diagram of the system.

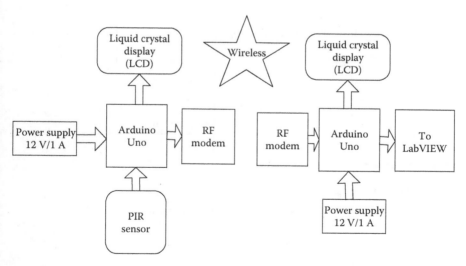

FIGURE 11.1
Block diagram for the system.

To design the system, following components are required (Tables 11.1 and 11.2).

TABLE 11.1

Components List for a Transmitter Section

Component/Specification	Quantity
Power supply/+12 V/1 A	1
Arduino Uno	1
PIR sensor	1
PIR sensor patch	1
LCD (20 * 4)	1
LCD patch	1
2.4 GHz RF modem	1
2.4 GHz RF modem patch	1
Connecting wires (M–M, M–F, F–F)	20 each
Zero size PCB or bread board or designed PCB	1

TABLE 11.2

Components List for a Receiver Section

Component/Specification	Quantity
Power supply/+12 V/1 A	1
Arduino Uno	1
LabVIEW software in PC/laptop	1
LCD (20 * 4)	1
LCD patch	1
2.4 GHz RF modem	1
2.4 GHz RF modem patch	1
Connecting wires (M–M, M–F, F–F)	20 each
Zero-size PCB or bread board or a designed PCB	1

11.2 Circuit Diagram

To develop the system, connect the components as follows:

TRANSMITTER SECTION

Arduino and PIR sensor

- Arduino GND-Sensor GND
- Arduino +5 V-sensor +Vcc
- Arduino A0 pin-data out pin of sensor

Arduino and RF modem

- Arduino pin0(Tx)-RX pin of RF modem
- Arduino pin0(RX)-Tx pin of RF modem
- Arduino+5 V-+Vcc of RF modem
- Arduino Ground-GND of RF modem

Arduino and LCD

- Arduino digital pin 13-RS pin(4) of LCD
- Arduino digital pin GND-RW pin(5) of LCD
- Arduino digital pin 12-E pin(6) of LCD
- Arduino digital pin 11-D4 pin(11) of LCD
- Arduino digital pin 10-D5 pin(12) of LCD
- Arduino digital pin 9-D6 pin(13) of LCD
- Arduino digital pin 8-D7 pin(14) of LCD

RECEIVER SECTION

Arduino and RF modem

- Arduino pin0(Tx)-RX pin of RF modem
- Arduino pin0(RX)-Tx pin of RF modem
- Arduino+5 V-+Vcc of RF modem
- Arduino Ground-GND of RF modem

Arduino and LCD

- Arduino digital pin 13-RS pin(4) of LCD
- Arduino digital pin GND-RW pin(5) of LCD
- Arduino digital pin 12-E pin(6) of LCD
- Arduino digital pin 11-D4 pin(11) of LCD
- Arduino digital pin 10-D5 pin(12) of LCD
- Arduino digital pin 9-D6 pin(13) of LCD
- Arduino digital pin 8-D7 pin(14) of LCD

Figure 11.2 shows the circuit diagram of the system.

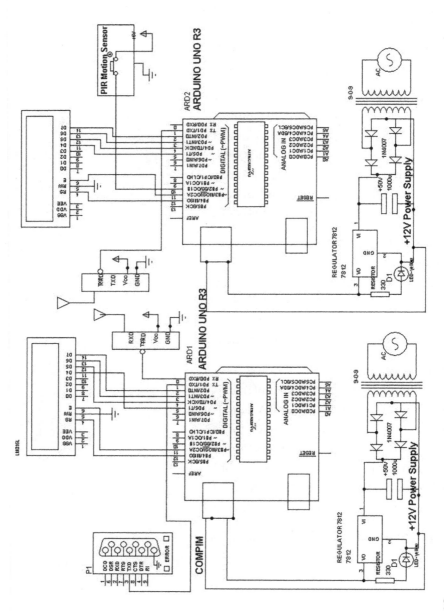

FIGURE 11.2
Circuit diagram.

11.3 Program

11.3.1 Transmitter Section

```
#include <LiquidCrystal.h>
LiquidCrystal lcd(12, 11, 5, 4, 3, 2);
const int PIR_sensor_PIN = 10;

int PIR_sensor_PIN_state=0;
void setup()
{
  lcd.begin(16,2);
  Serial.begin(9600);
  pinMode(PIR_sensor_PIN, INPUT);
  lcd.setCursor(0,0);
  lcd.print("SECURITY SYSTEM");
  lcd.setCursor(0,1);
  lcd.print("OF RESTRICTED AREA");
}

// the loop routine runs over and over again forever:
void loop()
{
  PIR_sensor_PIN_state = digitalRead(PIR_sensor_PIN);

  if(PIR_sensor_PIN_state == HIGH)
  {
    lcd.clear();
    int PIR_SERIAL_BYTE=10;
    Serial.write(PIR_SERIAL_BYTE);
    lcd.setCursor(0,1);
    lcd.print("SECURITY BREAK");
    delay(50);
  }

  else
  {
    lcd.clear();
    int PIR_SERIAL_BYTE=20;
    Serial.write(PIR_SERIAL_BYTE);
    lcd.setCursor(0,1);
    lcd.print("SYSTEM SECURED");
    delay(50);
  }

}
```

11.3.2 Receiver Section

```
#include <LiquidCrystal.h>
LiquidCrystal lcd(12, 11, 5, 4, 3, 2);

void setup()
{
  lcd.begin(16,2);
  Serial.begin(9600);
  lcd.setCursor(0,0);
  lcd.print("SECURITY SYSTEM");
  lcd.setCursor(0,1);
  lcd.print("OF RESTRICTED AREA");
}

// the loop routine runs over and over again forever:
void loop()
{
  int PIR_sensor_SERIAL_BYTE = Serial.read();;

  if(PIR_sensor_SERIAL_BYTE == 10)
  {
   lcd.clear();
   Serial.print("PIRSENSOR:");
   Serial.println(PIR_sensor_SERIAL_BYTE);
   lcd.setCursor(0,1);
   lcd.print("SECURITY BREAK");
   delay(50);
  }
  else if(PIR_sensor_SERIAL_BYTE == 20)
  {
   lcd.clear();
   Serial.print("PIRSENSOR:");
   Serial.println(PIR_sensor_SERIAL_BYTE);
   lcd.setCursor(0,1);
   lcd.print("SYSTEM SECURED");
   delay(50);
  }

}
```

11.4 Proteus Simulation Model

Connect all the components as described in Section 11.2. In addition to it, a logic needs to be connected with the sensor. As this is a virtual environment, so to make its input as LOW or HIGH, logic is connected to it, which can be changed manually to check the working of the sensor. Figure 11.3 shows a simulation model showing "SYSTEM SECURED" when logic is "0." Figure 11.4 shows a simulation model showing "SECURITY BREAK."

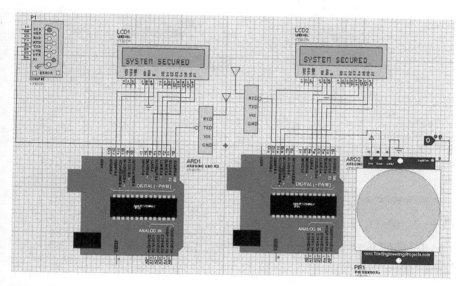

FIGURE 11.3
Proteus simulation model showing "system secured."

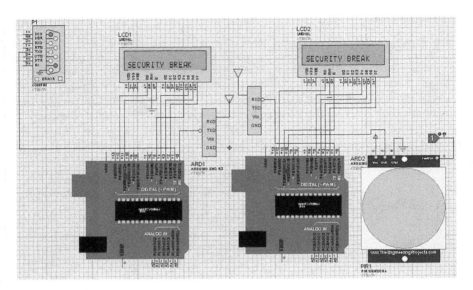

FIGURE 11.4
Proteus simulation model showing "security break."

11.5 LabVIEW GUI

LabVIEW GUI is designed for monitoring the security of the restricted areas. GUI is designed in two parts: (1) block diagram and (2) front panel. In front panel two LEDs are placed as an indicator when the system security is broken or the system is secured. When security is broken red LED glow and when the system is secured then the green LED glow. It has a serial COM port to access the data from the PC port, baud rate is set as "9600,"which is same as the wireless modem "Zigbee" to receive the data with the same rate, data bits "8," parity "none" stop bits "1," serial count is set as "100," and then Read_PIR_Serial to read the received data from the PIR sensor. Waveform chart is optional; it can be used to check the data received in the form of a waveform. To make front panel and block diagram, a graphical programming is done (Figures 11.5 through 11.7) with the help of the blocks and steps discussed in Chapter 4.

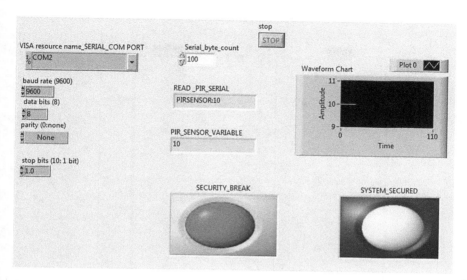

FIGURE 11.5
Front panel showing "security break."

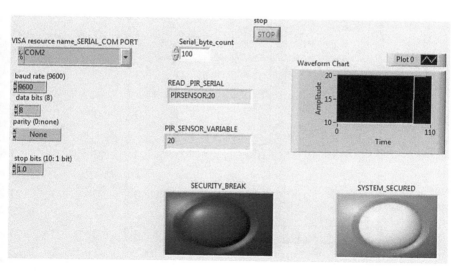

FIGURE 11.6
Front panel showing "system secured."

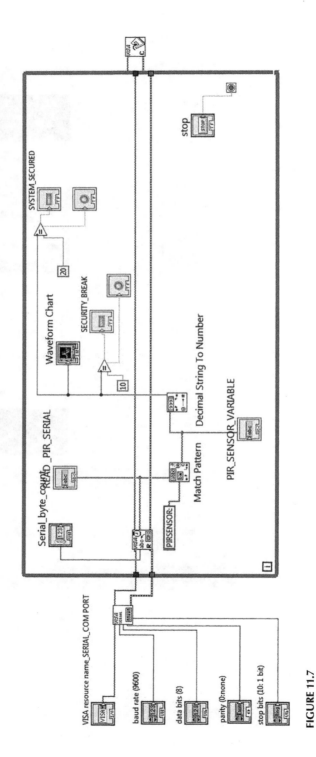

FIGURE 11.7
LabVIEW GUI block diagram for the system.

12

Campus Fire Monitoring System with a 2.4 GHz RF Modem

12.1 Introduction

The objective of this project is to design a monitoring system for the fire alerts in the campus. The system is designed in two sections: (1) transmitter section and (2) receiver section, in such a way that it will sense the fire in the surroundings with the help of a fire sensor at the transmitter section and communicated to the receiver section wirelessly with a 2.4 GHz RF modem. At the receiver end, a LabVIEW GUI is created to observe the system. Figure 12.1 shows the block diagram of the system.

To design the system, following components are required (Tables 12.1 and 12.2).

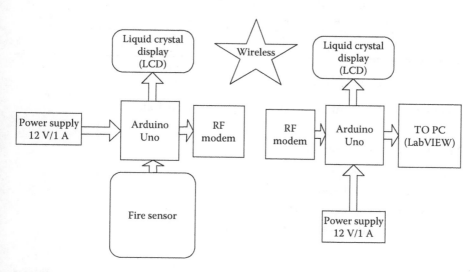

FIGURE 12.1
Block diagram for the system.

TABLE 12.1

Components List for a Transmitter Section

Component/Specification	Quantity
Power supply/+12 V/1 A	1
Arduino Uno	1
Flame sensor	1
Flame sensor patch	1
LCD (20 * 4)	1
LCD patch	1
2.4 GHz RF modem	1
2.4 GHz RF modem patch	1
Connecting wires (M–M, M–F, F–F)	20 each
Zero-size PCB or bread board or a designed PCB	1

TABLE 12.2

Components List for a Receiver Section

Component/Specification	Quantity
Power supply/+12 V/1 A	1
Arduino Uno	1
LabVIEW software in PC/laptop	1
LCD (20 * 4)	1
LCD patch	1
2.4 GHz RF modem	1
2.4 GHz RF modem patch	1
Connecting wires (M–M, M–F, F–F)	20 each
Zero-size PCB or bread board or a designed PCB	1

12.2 Circuit Diagram

To develop the system, connect the components as follows:

12.2.1 Transmitter Section

Arduino and FLAME sensor

- Arduino GND-Sensor GND
- Arduino +5 V-sensor +Vcc
- Arduino A0 pin-data out pin of sensor

Arduino and RF modem

- Arduino pin0(TX)-RX pin of RF modem
- Arduino pin0(RX)-Tx pin of RF modem
- Arduino+5 V-+Vcc of RF modem
- Arduino Ground-GND of RF modem

Arduino and LCD

- Arduino digital pin 13-RS pin(4) of LCD
- Arduino digital pin GND-RW pin(5) of LCD
- Arduino digital pin 12-E pin(6) of LCD
- Arduino digital pin 11-D4 pin(11) of LCD
- Arduino digital pin 10-D5 pin(12) of LCD
- Arduino digital pin 9-D6 pin(13) of LCD
- Arduino digital pin 8-D7 pin(14) of LCD

12.2.2 Receiver Section

Arduino and RF modem

- Arduino pin0(TX)-RX pin of RF modem
- Arduino pin0(RX)-Tx pin of RF modem
- Arduino+5 V-+Vcc of RF modem
- Arduino Ground-GND of RF modem

Arduino and LCD

- Arduino digital pin 13-RS pin(4) of LCD
- Arduino digital pin GND-RW pin(5) of LCD
- Arduino digital pin 12-E pin(6) of LCD
- Arduino digital pin 11-D4 pin(11) of LCD
- Arduino digital pin 10: D5 pin(12) of LCD
- Arduino digital pin 9-D6 pin(13) of LCD
- Arduino digital pin 8-D7 pin(14) of LCD

Figure 12.2 shows the circuit diagram for the system, connecting all the components as per connections.

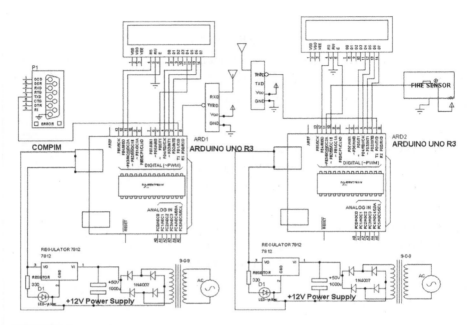

FIGURE 12.2
Circuit diagram.

12.3 Program

12.3.1 Transmitter Section

```
const int FLAME_SENSOR_PIN = 8;
int FLAME_SENSOR_state=0;
#include <LiquidCrystal.h>
LiquidCrystal lcd(12, 11, 5, 4, 3, 2);
void setup()
{
 lcd.begin(16,2);
 Serial.begin(9600);
 pinMode(FLAME_SENSOR_PIN, INPUT);
 lcd.setCursor(0,0);
 lcd.print("FLAME DETECTION");
 lcd.setCursor(0,1);
 lcd.print("SYSTENM AT UPES");
 delay(1000);

}
```

```
void loop()
{
 FLAME_SENSOR_state = digitalRead(FLAME_SENSOR_PIN);

 if(FLAME_SENSOR_state== HIGH)
 {
  lcd.clear();
  int FLAME_SENSOR_serial_data=10;
  Serial.write(FLAME_SENSOR_serial_data);
  lcd.setCursor(0,1);
  lcd.print("FLAME DETECTED");
  delay(50);
 }

 else
 {
  lcd.clear();
  int FLAME_SENSOR_serial_data=20;
  Serial.write(FLAME_SENSOR_serial_data);
  lcd.setCursor(0,1);
  lcd.print("NO FLAME");
  delay(50);
 }
}
```

12.3.2 Receiver Section

```
#include <LiquidCrystal.h>
LiquidCrystal lcd(12, 11, 5, 4, 3, 2);

void setup()
{
 lcd.begin(16,2);
 Serial.begin(9600);
 lcd.setCursor(0,0);
 lcd.print("FLAME DETECTION");
 lcd.setCursor(0,1);
 lcd.print("SYSTENM AT UPES");
 delay(1000);
}

void loop()
{
 int FLAME_SENSOR_serial_data= Serial.read();;
 if(FLAME_SENSOR_serial_data == 10)
 {
  lcd.clear();
  Serial.print("FLAMESENSOR:");
```

```
  Serial.println(FLAME_SENSOR_serial_data);
  lcd.setCursor(0,0);
  lcd.print("FLAME DETECTED");
  delay(50);
}
else if(FLAME_SENSOR_serial_data == 20)
{
  lcd.clear();
  Serial.print("FLAMESENSOR:");
  Serial.println(FLAME_SENSOR_serial_data);
  lcd.setCursor(0,1);
  lcd.print("NO FLAME");
  delay(50);
}
```

12.4 Proteus Simulation Model

Connect all the components as described in Section 12.2. In addition to it, a logic needs to be connected with the sensor. As this is a virtual environment, so to make its input as LOW or HIGH, a logic is connected to it, which can be changed manually to check the working of the sensor. Figure 12.3 shows a simulation model showing "NO FLAME" when logic is "0." Figure 12.4 shows simulation model showing "FLAME DETECTED."

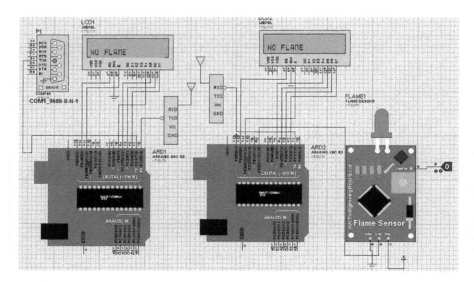

FIGURE 12.3
Proteus simulation model for the system showing "NO FLAME."

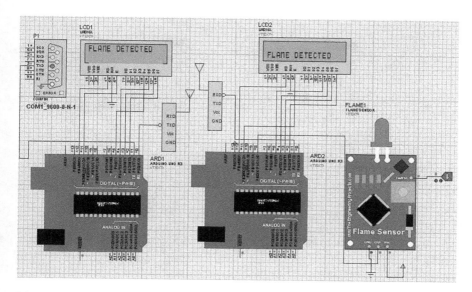

FIGURE 12.4
Proteus simulation model for the system showing "FLAME DETECTED."

12.5 LabVIEW GUI

LabVIEW GUI is designed for monitoring system for the fire alerts in the campus. GUI is designed in two parts: (1) block diagram and (2) front panel. In the front panel, two LEDs are placed as an indicator when system security is broken or the system is secured. Red LED indicates "FIRE DETECTED" and green LED indicates "NO FIRE DETECTED." It has a serial COM port to access the data from the PC port, baud rate is set as "9600," which is same as the wireless modem "Zigbee" to receive the data with the same rate, data bits "8," parity "none" stop bits "1," serial count is set as "100," flow control "None" Serial_read_buffer to read the received data from the fire sensor. Waveform chart is optional; it can be used to check the data received in form of waveform. To make the front panel and block diagram, active graphical programming is done (Figures 12.5 through 12.7) with the help of blocks and steps discussed in Chapter 4.

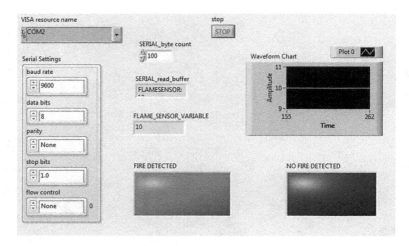

FIGURE 12.5
Front panel showing "FIRE DETECTED."

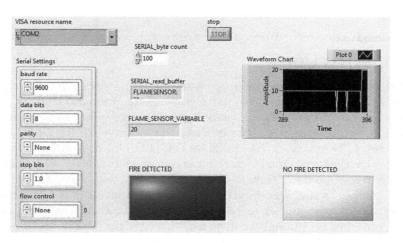

FIGURE 12.6
Front panel showing "NO FIRE DETECTED."

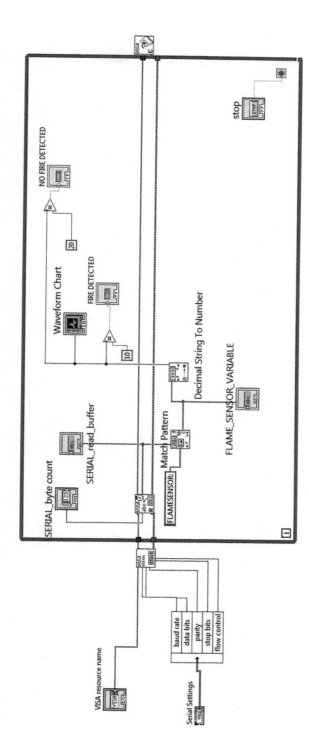

FIGURE 12.7
LabVIEW GUI block diagram for the system.

13

Light-Dependent Resistor-Based Light Intensity Control System

13.1 Introduction

This project is about to design the light-dependent resistor-based room light intensity control system. The system is designed in two sections: (1) transmitter section and (2) receiver section. At the transmitter, light-dependent resistor (LDR) is connected to the Arduino to sense the intensity of the surroundings. The level of intensity is sensed and transmitted to the receiver section to control the intensity at the required level. The receiver section comprises the Arduino Uno, LCD, RF modem, dimmer, and load (bulb). The load can be controlled in 4, 8, or 16 levels. At the server, a LabVIEW GUI is created to observe the system. Figure 13.1 shows the block diagram of the system.

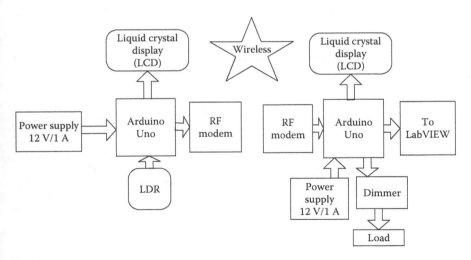

FIGURE 13.1
Block diagram for the system.

To design the system, following components are required (Tables 13.1 and 13.2).

TABLE 13.1

Components List for a Transmitter Section

Component/Specification	Quantity
Power supply/+12 V/1 A	1
Arduino Uno	1
LDR sensor	1
LDR sensor patch	1
LCD (20 * 4)	1
LCD patch	1
2.4 GHz RF modem	1
2.4 GHz RF modem patch	1
Connecting wires (M–M, M–F, F–F)	20 each
Zero-size PCB or bread board or a designed PCB	1

TABLE 13.2

Components List for a Receiver Section

Component/Specification	Quantity
Power supply/+12 V/1 A	1
Arduino Uno	1
LabVIEW software in PC/laptop	1
LCD (20 * 4)	1
LCD patch	1
Dimmer	1
Dimmer patch	1
2.4 GHz RF modem	1
2.4 GHz RF modem patch	1
Connecting wires (M–M, M–F, F–F)	20 each
Zero-size PCB or bread board or a designed PCB	1

13.2 Circuit Diagram

To develop the system, connect the components as follows:

13.2.1 Transmitter Section

Arduino and LDR sensor

- Arduino GND-Sensor GND
- Arduino +5 V-sensor +Vcc
- Arduino A0 pin-data out pin of sensor

Arduino and RF modem

- Arduino pin0(TX)-RX pin of RF modem
- Arduino pin0(RX)-Tx pin of RF modem
- Arduino+5 V-+Vcc of RF modem
- Arduino Ground-GND of RF modem

Arduino and LCD

- Arduino digital pin 13-RS pin(4) of LCD
- Arduino digital pin GND-RW pin(5) of LCD
- Arduino digital pin 12-E pin(6) of LCD
- Arduino digital pin 11-D4 pin(11) of LCD
- Arduino digital pin 10-D5 pin(12) of LCD
- Arduino digital pin 9-D6 pin(13) of LCD
- Arduino digital pin 8-D7 pin(14) of LCD

13.2.2 Receiver Section

Arduino and RF modem

- Arduino pin0(TX)-RX pin of RF modem
- Arduino pin0(RX)-Tx pin of RF modem
- Arduino+5 V-+Vcc of RF modem
- Arduino Ground-GND of RF modem

Arduino and LCD

- Arduino digital pin 13-RS pin(4) of LCD
- Arduino digital pin GND-RW pin(5) of LCD
- Arduino digital pin 12-E pin(6) of LCD
- Arduino digital pin 11-D4 pin(11) of LCD
- Arduino digital pin 10-D5 pin(12) of LCD
- Arduino digital pin 9-D6 pin(13) of LCD
- Arduino digital pin 8-D7 pin(14) of LCD

Figure 13.2 shows the circuit diagram for the system.

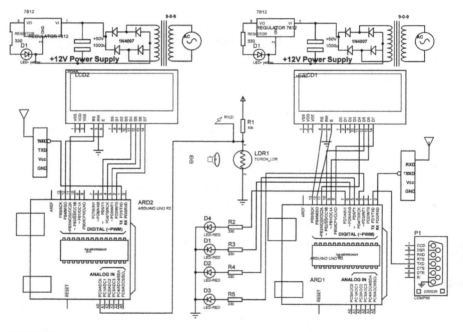

FIGURE 13.2
Circuit diagram for the system.

13.3 Program

13.3.1 Transmitter Section

```
#include <LiquidCrystal.h>
LiquidCrystal lcd(12, 11, 5, 4, 3, 2);
```

```
#define LDR_SENSOR_PIN A0
void setup()
{
 lcd.begin(20,4);
 Serial.begin(9600);
 lcd.setCursor(0,0);
 lcd.print( "LDR BASED LIGHT");
 lcd.setCursor(0,1);
 lcd.print( "Intensity control ");
 lcd.setCursor(0,2);
 lcd.print( "system AT SME ");
 delay(1000);
}

void loop()
{
 int LDR_sensor_BYTE_LEVEL = analogRead(LDR_SENSOR_PIN);
 int LDR_sensor_BYTE=LDR_sensor_BYTE_LEVEL/100;
 lcd.setCursor(0,3);
 lcd.print("LDR LEVEL:");
 lcd.setCursor(11,3);
 lcd.print(LDR_sensor_BYTE);
 Serial.write(LDR_sensor_BYTE);
 delay(100);
}
```

13.3.2 Receiver Section

```
#include <LiquidCrystal.h>
LiquidCrystal lcd(13, 12, 11, 10, 9,8);
#define DIMMER_PIN_A 7
#define DIMMER_PIN_B 6
#define DIMMER_PIN_C 5
#define DIMMER_PIN_D 4
void setup()
{
 lcd.begin(20,4);
 Serial.begin(9600);
 lcd.setCursor(0,0);
 lcd.print( "LDR BASED LIGHT");
 lcd.setCursor(0,1);
 lcd.print( "Intensity control ");
 lcd.setCursor(0,2);
 lcd.print( "system AT SME ");
 pinMode(DIMMER_PIN_A,OUTPUT);
 pinMode(DIMMER_PIN_B,OUTPUT);
 pinMode(DIMMER_PIN_C,OUTPUT);
 pinMode(DIMMER_PIN_D,OUTPUT);
 delay(1000);
}
```

```
void loop()
{
 int LDR_sensor_BYTE=Serial.read();

 Serial.print("LDRSENSOR:");
 Serial.println(LDR_sensor_BYTE);
  lcd.setCursor(0,3);
  lcd.print("LDR LEVEL:");
  lcd.setCursor(11,3);
  lcd.print(LDR_sensor_BYTE);

  if(LDR_sensor_BYTE<=20)
  {
   digitalWrite(DIMMER_PIN_A,LOW);
   digitalWrite(DIMMER_PIN_B,LOW);
   digitalWrite(DIMMER_PIN_C,LOW);
   digitalWrite(DIMMER_PIN_D,HIGH);
  }
  if(LDR_sensor_BYTE>=30)
  {
   digitalWrite(DIMMER_PIN_A,LOW);
   digitalWrite(DIMMER_PIN_B,LOW);
   digitalWrite(DIMMER_PIN_C,HIGH);
   digitalWrite(DIMMER_PIN_D,LOW);
  }

  delay(100);
}
```

13.4 Proteus Simulation Model

Proteus Simulation model for the system is designed by connecting all the components as described in Section 13.2. Figure 13.3 shows a Proteus simulation model with LDR levels.

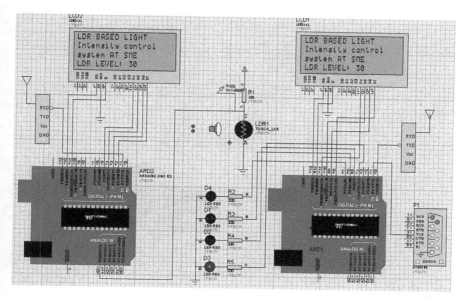

FIGURE 13.3
Proteus simulation model for the system.

13.5 LabVIEW GUI

LabVIEW GUI is designed for LDR-based light-intensity control system. GUI is designed in two parts: (1) block diagram and the (2) front panel. The front panel has a serial COM port to access the data from the PC port, baud rate is set as "9600," which is same as the wireless modem "Zigbee" to receive the data with the same rate, data bits "8," parity "none" stop bits "1," serial count is set as "100," flow control "None," and Serial_read_buffer to read the received data from the fire sensor. Waveform chart is optional; it can be used to check the data received in the form of waveform. To make the front panel, an active graphical programming is done in the block diagram (Figures 13.4 through 13.5) with the help of blocks and steps discussed in Chapter 4.

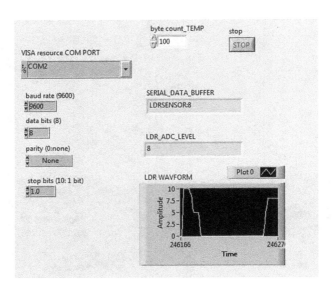

FIGURE 13.4
Front panel for the system.

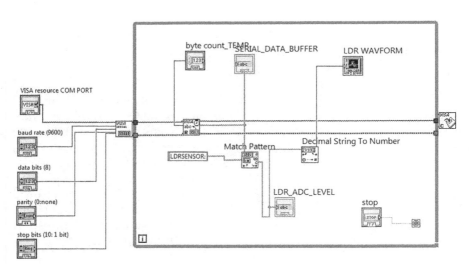

FIGURE 13.5
LabVIEW GUI block diagram for the system.

14

DC Motor Control System
with LabVIEW GUI

14.1 Introduction

This project is designed to control DC motor with the LabVIEW GUI. It can be used in the applications such as a wireless control for a robot. The system is designed in two sections: (1) transmitter section and the (2) receiver section. The transmitter section comprises Arduino Uno, power supply, PC (for LabVIEW), LCD, and the RF modem. In this project, two independent methods to control a DC motor are discussed. A motor can be controlled with a switch array as well as with LabVIEW GUI for "clockwise," "anticlockwise," and "stop" to make movement in the forward or reverse direction. Receiver section comprises the Arduino Uno, power supply, motor driver (L293D), DC motor, and RF modem. The RF modem is connected to communicate wirelessly between the two sections and operates at 2.4 GHz frequency. Figure 14.1 shows the block diagram of the system.

To design the system, following components are required (Tables 14.1 and 14.2).

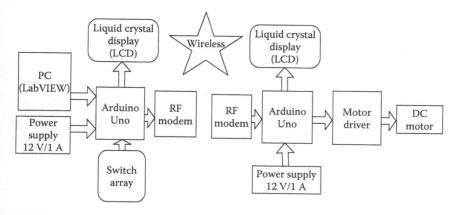

FIGURE 14.1
Block diagram for the system.

TABLE 14.1

Components List for a Transmitter Section

Component/Specification	Quantity
Power supply/+12 V/1 A	1
Power supply/+5 V/500 mA	1
Arduino Uno	1
Switches	4
PC/laptop (LabVIEW)	1
LCD (20 * 4)	1
LCD patch	1
2.4 GHz RF modem	1
2.4 GHz RF modem patch	1
Connecting wires (M–M, M–F, F–F)	20 each
Zero-size PCB or bread board or a designed PCB	1

TABLE 14.2

Components List for a Receiver Section

Component/Specification	Quantity
Power supply/+12 V/1 A	1
Arduino Uno	1
DC motor	1
L293D	1
LCD (20 * 4)	1
LCD patch	1
2.4 GHz RF modem	1
2.4 GHz RF modem patch	1
Connecting wires (M–M, M–F, F–F)	20 each
Zero-size PCB or bread board or a designed PCB	1

14.2 Circuit Diagram

To develop the system, connect the components as follows:

14.2.1 Transmitter Section

Arduino and push button

- Arduino pin7-one terminal of push button 'clockwise'
- Arduino pin6-one terminal of push button 'anti-clockwise'
- Arduino pin5-one terminal of push button 'stop'
- Other terminal of all push buttons-GND

Arduino and RF modem

- Arduino pin0(TX)-RX pin of RF modem
- Arduino pin0(RX)-Tx pin of RF modem
- Arduino+5 V-+Vcc of RF modem
- Arduino Ground-GND of RF modem

Arduino and LCD

- Arduino digital pin 13-RS pin(4) of LCD
- Arduino digital pin GND-RW pin(5) of LCD
- Arduino digital pin 12-E pin(6) of LCD
- Arduino digital pin 11-D4 pin(11) of LCD
- Arduino digital pin 10-D5 pin(12) of LCD
- Arduino digital pin 9-D6 pin(13) of LCD
- Arduino digital pin 8-D7 pin(14) of LCD

14.2.2 Receiver Section

Arduino and RF modem

- Arduino pin0(TX)-RX pin of RF modem
- Arduino pin0(RX)-Tx pin of RF modem
- Arduino+5 V-+Vcc of RF modem
- Arduino Ground-GND of RF modem

Arduino and LCD

- Arduino digital pin 13-RS pin(4) of LCD
- Arduino digital pin GND-RW pin(5) of LCD
- Arduino digital pin 12-E pin(6) of LCD
- Arduino digital pin 11-D4 pin(11) of LCD
- Arduino digital pin 10-D5 pin(12) of LCD
- Arduino digital pin 9-D6 pin(13) of LCD
- Arduino digital pin 8-D7 pin(14) of LCD

Arduino and L293D connection

- Arduino GND-4,5,12,13 pins of IC
- Arduino +5 V-1,9,16 pins of IC
- Arduino pin 7-pin 2 of IC
- Arduino pin 6-pin 7 of IC
- Arduino pin 5-pin 10 of IC
- Arduino pin 4-pin 15 of IC
- L293D pin8-+ve of 12 V battery

L293D and DC motor connection

- L293D pin 3-to +ve pin of DC motor1
- L293D pin 6-to −ve pin of DC motor1

Figure 14.2 shows the circuit diagram for the system.

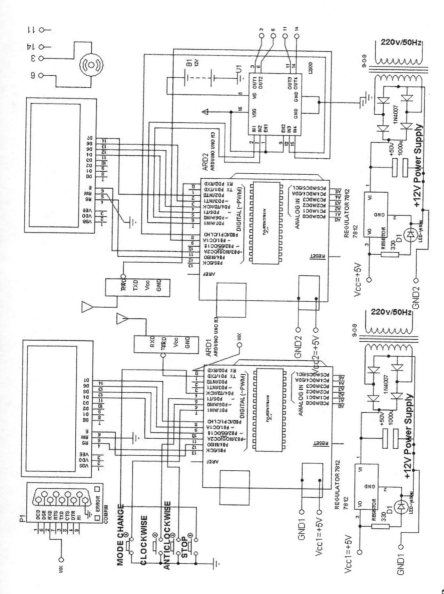

FIGURE 14.2
Circuit diagram.

14.3 Program

14.3.1 Transmitter Section

```
#include <LiquidCrystal.h>
LiquidCrystal lcd(13, 12, 11, 10, 9, 8);
#define BUTTON_MODE_SELECTION 7
#define BUTTON_PIN_FORWARD 6
#define BUTTON_PIN_REVERSE 5
#define BUTTON_PIN_STOP 4
void setup()
{
 lcd.begin(20,4);
 Serial.begin(9600);
 pinMode(BUTTON_MODE_SELECTION,INPUT_PULLUP);
 pinMode(BUTTON_PIN_FORWARD,INPUT_PULLUP);
 pinMode(BUTTON_PIN_REVERSE,INPUT_PULLUP);
 pinMode(BUTTON_PIN_STOP,INPUT_PULLUP);
 lcd.setCursor(0,0);
 lcd.print("DC motor control");
 lcd.setCursor(0,1);
 lcd.print("+ remote control");
 delay(1000);
}

void loop()
{
 int
 BUTTON_MODE_SELECTION_digital=digitalRead(BUTTON_MODE_
 SELECTION);
 if (BUTTON_MODE_SELECTION_digital==LOW)
  {
   char LABVIEW_SERIAL_CHAR;
   LABVIEW_SERIAL_CHAR=Serial.read();
    if (LABVIEW_SERIAL_CHAR=='W')

    {
     lcd.clear();
     lcd.setCursor(0,3);
     lcd.print("CLOSEWISE");
     Serial.write('W');

    }


    else if (LABVIEW_SERIAL_CHAR=='Z')
    {
     lcd.clear();
```

```
      lcd.setCursor(0,3);
      lcd.print("ANTICLOCKWISE");
      Serial.write('Z');

     }


    else if (LABVIEW_SERIAL_CHAR=='E')
    {
     lcd.clear();
     lcd.setCursor(0,3);
     lcd.print("STOP");
     Serial.write('E');
    }
    delay(10);
   }

 else if(BUTTON_MODE_SELECTION_digital==HIGH)
  {
   int
   BUTTON_PIN_FORWARD_digital=digitalRead (BUTTON_PIN_FORWARD);
   int BUTTON_PIN_REVERSE_digital=digitalRead
   (BUTTON_PIN_REVERSE);
   int BUTTON_PIN_STOP_digital=digitalRead (BUTTON_PIN_STOP);
   if (BUTTON_PIN_FORWARD_digital==LOW)
    {
     lcd.clear();
     lcd.setCursor(0,3);
     lcd.print("CLOSEWISE");
     Serial.write('W');
    }
   if(BUTTON_PIN_REVERSE_digital==LOW)
    {
     lcd.clear();
     lcd.setCursor(0,3);
     lcd.print("ANTICLOCKWISE");
     Serial.write('Z');
    }
   if(BUTTON_PIN_STOP_digital==LOW)
    {
     lcd.clear();
     lcd.setCursor(0,3);
     lcd.print("STOP");
     Serial.write('E');
    }

  }

}
```

14.3.2 Receiver Program

```
#include <LiquidCrystal.h>
LiquidCrystal lcd(12, 11, 5, 4, 3, 2);
#define DC_MOTOR1_POSITIVE 6
#define DC_MOTOR1_NEGATIVE 7

void setup()
{
 Serial.begin(9600);
 lcd.begin(20,4);
 pinMode(DC_MOTOR1_POSITIVE, OUTPUT);
 pinMode(DC_MOTOR1_NEGATIVE, OUTPUT);
 lcd.setCursor(0,0);
 lcd.print("DC motor control");
 lcd.setCursor(0,1);
 lcd.print("+ remote control");
 delay(1000);
}

void loop()
{
char LABVIEW_SERIAL_CHAR;
LABVIEW_SERIAL_CHAR=Serial.read();
 if (LABVIEW_SERIAL_CHAR=='W')

  {
   lcd.clear();
   lcd.setCursor(0,3);
   lcd.print("CLOCKWISE");
   digitalWrite(DC_MOTOR1_POSITIVE, HIGH);
   digitalWrite(DC_MOTOR1_NEGATIVE, LOW);
  }
```

```
else if (LABVIEW_SERIAL_CHAR=='Z')
{
 lcd.clear();
 lcd.setCursor(0,3);
 lcd.print("ANTICLOCKWISE");
 digitalWrite(DC_MOTOR1_POSITIVE, LOW);
 digitalWrite(DC_MOTOR1_NEGATIVE, HIGH);
}

else if (LABVIEW_SERIAL_CHAR=='E')
{
  lcd.clear();
  lcd.setCursor(0,3);
  lcd.print("STOP");
   digitalWrite(DC_MOTOR1_POSITIVE, LOW);
   digitalWrite(DC_MOTOR1_NEGATIVE, LOW);

}

 delay(10);
}
```

14.4 Proteus Simulation Model

Connect all the components as described in Section 14.2. Figure 14.3 shows a simulation model showing that the motor is moving "CLOCKWISE."

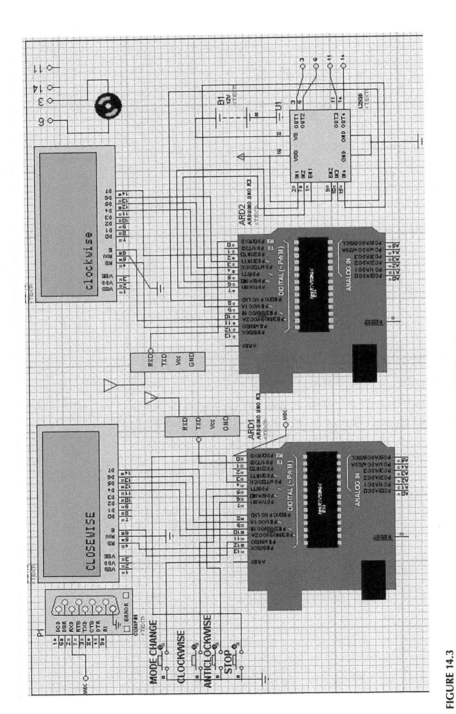

FIGURE 14.3
Proteus simulation model showing the motor moving in "CLOCKWISE" direction.

14.5 LabVIEW GUI

LabVIEW GUI is designed to control the DC motor. GUI is designed in two parts: (1) block diagram and (2) front panel. In the front panel, three buttons are used to make the motor in "forward," "reverse," or "stop." It has a serial COM port to access the data from the PC port, baud rate is set as "9600" which is same as the wireless modem "Zigbee" to send the data with same rate, data bits "8," parity "none" stop bits "1." To make the front panel, an active graphical programming is done in the block diagram (Figures 14.4 through 14.6) with the help of blocks and steps discussed in Chapter 4.

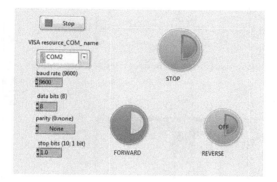

FIGURE 14.4
Front panel showing that the motor is controlled in "FORWARD" direction.

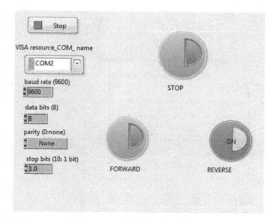

FIGURE 14.5
Front panel showing that the motor is controlled in "REVERSE" direction.

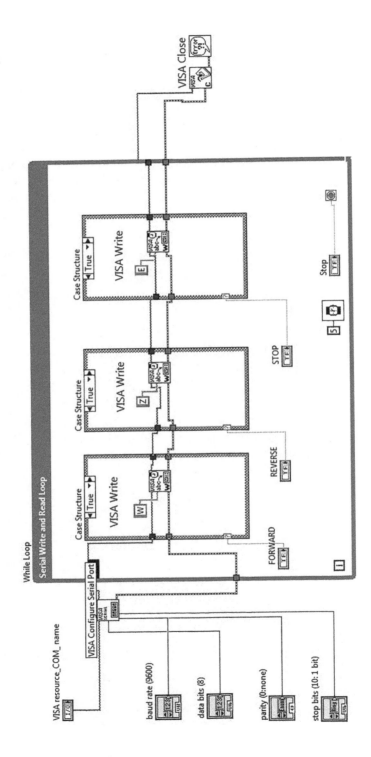

FIGURE 14.6
LabVIEW GUI block diagram for the system.

15

Stepper Motor Control System with LabVIEW GUI

15.1 Introduction

This project is designed to control the stepper motor. The system is designed to operate with the help of LabVIEW GUI. It comprises of two sections: (1) transmitter section and (2) receiver section. Transmitter section comprises the Arduino Uno, power supply, PC (for LabVIEW), LCD, and the RF modem. Motor is controlled with LabVIEW GUI. Receiver section comprises the Arduino Uno, power supply, motor driver (L293D), stepper motor, and the RF modem. The RF modem is connected to communicate wirelessly between the two sections and operates at 2.4 GHz frequency. Figure 15.1 shows the block diagram of the system.

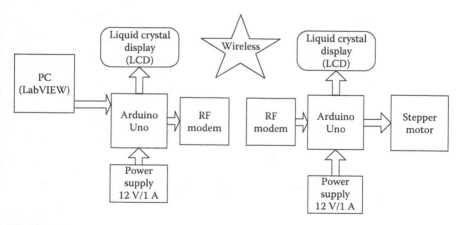

FIGURE 15.1
Block diagram for the system.

To design the system, following components are required (Tables 15.1 and 15.2)

TABLE 15.1

Components List for a Transmitter Section

Component/Specification	Quantity
Power supply/+12 V/1 A	1
Power supply/+5 V/500 mA	1
Arduino Uno	1
PC/Laptop (LabVIEW)	1
LCD (20 * 4)	1
LCD patch	1
2.4 GHz RF modem	1
2.4 GHz RF modem patch	1
Connecting wires (M–M, M–F, F–F)	20 each
Zero-size PCB or bread board or a designed PCB	1

TABLE 15.2

Components List for a Receiver Section

Component/Specification	Quantity
Power supply/+12 V/1 A	1
Power supply/+5 V/500 mA	1
Arduino Uno	1
LCD (20 * 4)	1
LCD patch	1
Stepper motor	1
L293D	1
2.4 GHz RF modem	1
2.4 GHz RF modem patch	1
Connecting wires (M–M, M–F, F–F)	20 each
Zero-size PCB or bread board or a designed PCB	1

15.2 Circuit Diagram

To develop the system, connect the components as follows:

15.2.1 Transmitter Section

Arduino and push button

- Arduino pin7-push button clockwise
- Arduino pin6-push button anticlockwise
- Arduino pin5-push button stop

Arduino and RF modem

- Arduino pin0(TX)-RX pin of RF modem
- Arduino pin0(RX)-Tx pin of RF modem
- Arduino+5 V-+Vcc of RF modem
- Arduino Ground-GND of RF modem

Arduino and LCD

- Arduino digital pin 13-RS pin(4) of LCD
- Arduino digital pin GND-RW pin(5) of LCD
- Arduino digital pin 12-E pin(6) of LCD
- Arduino digital pin 11-D4 pin(11) of LCD
- Arduino digital pin 10-D5 pin(12) of LCD
- Arduino digital pin 9-D6 pin(13) of LCD
- Arduino digital pin 8-D7 pin(14) of LCD

15.2.2 Receiver Section

Arduino and RF modem

- Arduino pin0(TX)-RX pin of RF modem
- Arduino pin0(RX)-Tx pin of RF modem
- Arduino+5 V-+Vcc of RF modem
- Arduino Ground-GND of RF modem

Arduino and LCD

- Arduino digital pin 13-RS pin(4) of LCD
- Arduino digital pin GND-RW pin(5) of LCD
- Arduino digital pin 12-E pin(6) of LCD
- Arduino digital pin 11-D4 pin(11) of LCD
- Arduino digital pin 10-D5 pin(12) of LCD
- Arduino digital pin 9-D6 pin(13) of LCD
- Arduino digital pin 8-D7 pin(14) of LCD

Arduino and L293D connection

- Arduino GND-4,5,12,13 pins of IC
- Arduino +5 V-1,9,16 pins of IC
- Arduino pin 7-pin 2 of IC
- Arduino pin 6-pin 7 of IC
- Arduino pin 5-pin 10 of IC
- Arduino pin 4-pin 15 of IC
- L293D pin8-+ve of 12 V battery

L293D and Stepper motor connection

- L293D pin 3-A phase of stepper motor
- L293D pin 6-B phase of stepper motor
- L293D pin 11-C phase of stepper motor
- L293D pin 14-D phase of stepper motor

Figure 15.2 shows the circuit diagram for the system, connecting all the components as per connections.

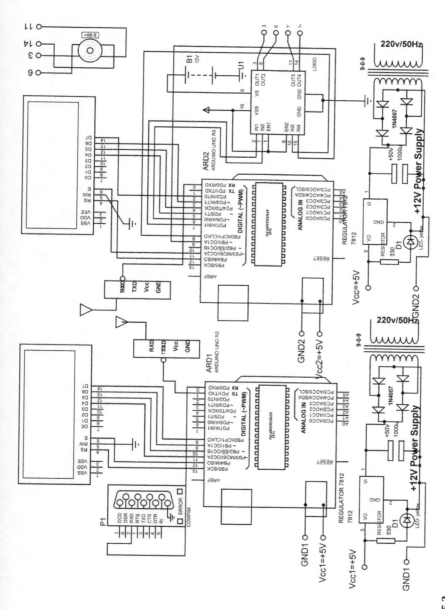

FIGURE 15.2
Circuit diagram.

15.3 Program

15.3.1 Transmitter Section

```
#include <LiquidCrystal.h>
LiquidCrystal lcd(13, 12, 11, 10, 9, 8);

void setup()
{
 lcd.begin(20,4);
 Serial.begin(9600);
 lcd.setCursor(0,0);
 lcd.print("DC motor control");
 lcd.setCursor(0,1);
 lcd.print("+ remote control");
 delay(1000);
}

void loop()
{
char LABVIEW_SERIAL_CHAR;
LABVIEW_SERIAL_CHAR=Serial.read();
 if (LABVIEW_SERIAL_CHAR=='W')

 {
   lcd.clear();
   lcd.setCursor(0,3);
   lcd.print("4 STEP SYSTEM");
   Serial.write('W');

 }
 else if (LABVIEW_SERIAL_CHAR=='Z')
 {
   lcd.clear();
   lcd.setCursor(0,3);
   lcd.print("8 STEP SYSTEM");
   Serial.write('Z');

 }

 else if (LABVIEW_SERIAL_CHAR=='E')
 {
   lcd.clear();
   lcd.setCursor(0,3);
   lcd.print("STOP");
   Serial.write('E');
 }
  delay(10);
}
```

15.3.2 Receiver Program

```
#include <LiquidCrystal.h>
LiquidCrystal lcd(12, 11, 5, 4, 3, 2);
#define STEPPER_MOTOR_PHASE_A 6
#define STEPPER_MOTOR_PHASE_B 7
#define STEPPER_MOTOR_PHASE_C 8
#define STEPPER_MOTOR_PHASE_D 9
void setup()
{
 Serial.begin(9600);
 lcd.begin(20,4);
 pinMode(STEPPER_MOTOR_PHASE_A, OUTPUT);
 pinMode(STEPPER_MOTOR_PHASE_B, OUTPUT);
 pinMode(STEPPER_MOTOR_PHASE_C, OUTPUT);
 pinMode(STEPPER_MOTOR_PHASE_D, OUTPUT);
 lcd.setCursor(0,0);
 lcd.print("DC motor control");
 lcd.setCursor(0,1);
 lcd.print("+ remote control");
 delay(1000);
}

void loop()
{
char LABVIEW_SERIAL_CHAR;
LABVIEW_SERIAL_CHAR=Serial.read();
 if (LABVIEW_SERIAL_CHAR=='W')

 {
    lcd.clear();
    lcd.setCursor(0,3);
    lcd.print("4 STEP SYSTEM");
    digitalWrite(STEPPER_MOTOR_PHASE_A, HIGH);
    digitalWrite(STEPPER_MOTOR_PHASE_B, LOW);
    digitalWrite(STEPPER_MOTOR_PHASE_C, LOW);
    digitalWrite(STEPPER_MOTOR_PHASE_D, LOW);
    delay(20);
    digitalWrite(STEPPER_MOTOR_PHASE_A, LOW);
    digitalWrite(STEPPER_MOTOR_PHASE_B, HIGH);
    digitalWrite(STEPPER_MOTOR_PHASE_C, LOW);
    digitalWrite(STEPPER_MOTOR_PHASE_D, LOW);
    delay(20);
    digitalWrite(STEPPER_MOTOR_PHASE_A, LOW);
    digitalWrite(STEPPER_MOTOR_PHASE_B, LOW);
    digitalWrite(STEPPER_MOTOR_PHASE_C, HIGH);
    digitalWrite(STEPPER_MOTOR_PHASE_D, LOW);
    delay(20);
    digitalWrite(STEPPER_MOTOR_PHASE_A, LOW);
    digitalWrite(STEPPER_MOTOR_PHASE_B, LOW);
```

```
     digitalWrite(STEPPER_MOTOR_PHASE_C, LOW);
     digitalWrite(STEPPER_MOTOR_PHASE_D, HIGH);
}
else if (LABVIEW_SERIAL_CHAR=='Z')
{
  lcd.clear();
  lcd.setCursor(0,3);
  lcd.print("Motor Reverse");
  digitalWrite(STEPPER_MOTOR_PHASE_A, HIGH);
  digitalWrite(STEPPER_MOTOR_PHASE_B, LOW);
  digitalWrite(STEPPER_MOTOR_PHASE_C, LOW);
  digitalWrite(STEPPER_MOTOR_PHASE_D, LOW);
  delay(20);
  digitalWrite(STEPPER_MOTOR_PHASE_A, HIGH);
  digitalWrite(STEPPER_MOTOR_PHASE_B, HIGH);
  digitalWrite(STEPPER_MOTOR_PHASE_C, LOW);
  digitalWrite(STEPPER_MOTOR_PHASE_D, LOW);
  delay(20);
  digitalWrite(STEPPER_MOTOR_PHASE_A, LOW);
  digitalWrite(STEPPER_MOTOR_PHASE_B, HIGH);
  digitalWrite(STEPPER_MOTOR_PHASE_C, LOW);
  digitalWrite(STEPPER_MOTOR_PHASE_D, LOW);
  delay(20);
  digitalWrite(STEPPER_MOTOR_PHASE_A, LOW);
  digitalWrite(STEPPER_MOTOR_PHASE_B, HIGH);
  digitalWrite(STEPPER_MOTOR_PHASE_C, HIGH);
  digitalWrite(STEPPER_MOTOR_PHASE_D, LOW);
  delay(20);
  digitalWrite(STEPPER_MOTOR_PHASE_A, LOW);
  digitalWrite(STEPPER_MOTOR_PHASE_B, LOW);
  digitalWrite(STEPPER_MOTOR_PHASE_C, HIGH);
  digitalWrite(STEPPER_MOTOR_PHASE_D, LOW);
  delay(20);
  digitalWrite(STEPPER_MOTOR_PHASE_A, LOW);
  digitalWrite(STEPPER_MOTOR_PHASE_B, LOW);
  digitalWrite(STEPPER_MOTOR_PHASE_C, HIGH);
  digitalWrite(STEPPER_MOTOR_PHASE_D, HIGH);
  delay(20);
  digitalWrite(STEPPER_MOTOR_PHASE_A, LOW);
  digitalWrite(STEPPER_MOTOR_PHASE_B, LOW);
  digitalWrite(STEPPER_MOTOR_PHASE_C, LOW);
  digitalWrite(STEPPER_MOTOR_PHASE_D,HIGH);
  delay(20);
```

```
   digitalWrite(STEPPER_MOTOR_PHASE_A, HIGH);
   digitalWrite(STEPPER_MOTOR_PHASE_B, LOW);
   digitalWrite(STEPPER_MOTOR_PHASE_C, LOW);
   digitalWrite(STEPPER_MOTOR_PHASE_D, HIGH);

}

 else if (LABVIEW_SERIAL_CHAR=='E')
 {
   lcd.clear();
   lcd.setCursor(0,3);
   lcd.print("STOP");
    digitalWrite(STEPPER_MOTOR_PHASE_A, LOW);
    digitalWrite(STEPPER_MOTOR_PHASE_B, LOW);
    digitalWrite(STEPPER_MOTOR_PHASE_C, LOW);
    digitalWrite(STEPPER_MOTOR_PHASE_D, LOW);
 }
    delay(10);
}
```

15.4 Proteus Simulation Model

Connect all the components as described in Section 15.2. Figure 15.3 shows simulation model to control the stepper motor.

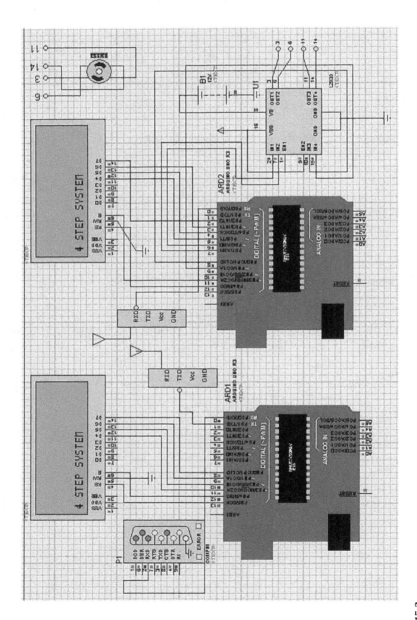

FIGURE 15.3
Proteus simulation model to control stepper motor.

15.5 LabVIEW GUI

LabVIEW GUI is designed to the control stepper motor. GUI is designed in two parts: (1) block diagram and (2) front panel. In the front panel, three buttons are used to make the motor in "forward," "reverse," and "stop" with 90° of the step. It has a serial COM port to access the data from PC port, baud rate is set as "9600," which is same as the wireless modem "Zigbee" to send the data with the same rate, data bits "8," parity "none," stop bits "1." To make the front panel, an active graphical programming is done in the block diagram (Figures 15.4 through 15.6) with the help of blocks and steps discussed in Chapter 4.

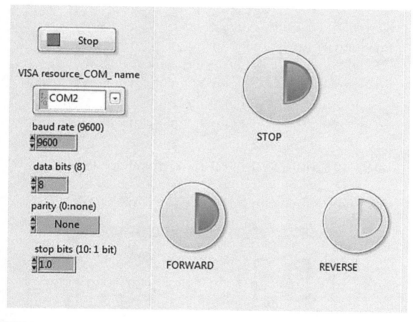

FIGURE 15.4
Front panel showing that the motor is controlled in the "REVERSE" direction.

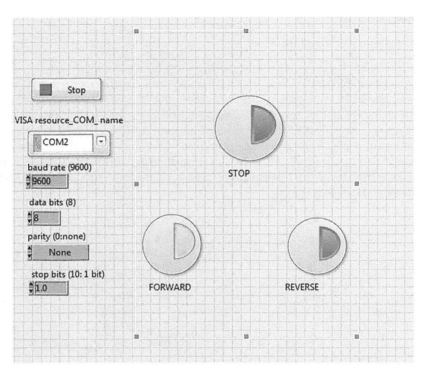

FIGURE 15.5
Front panel showing that the motor is controlled in the "FORWARD" direction.

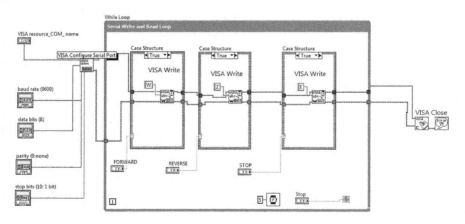

FIGURE 15.6
LabVIEW GUI block diagram for the system.

16

Accelerometer-Based Laboratory Automation System

16.1 Introduction

This project is to design an accelerometer-based laboratory automation system. The system is designed in two sections: (1) transmitter section and (2) receiver section. The transmitter section comprises the Arduino Nano, power supply, PC (for LabVIEW), Accelerometer, LCD, and the RF modem. The receiver section comprises the Arduino Nano, power supply, relay, load, and the RF modem. The RF modem is connected to communicate wirelessly between the two sections and operates at 2.4 GHz frequency. Load is made "ON" or "OFF" through relays. For the experiment, two load bulbs and fan are considered to operate. Figure 16.1 shows the block diagram of the system.

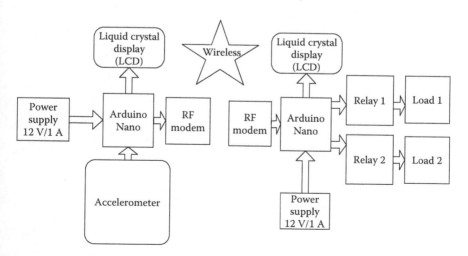

FIGURE 16.1
Block diagram for the system.

To design the system following components are required (Tables 16.1 and 16.2).

TABLE 16.1

Components List for a Transmitter Section

Component/Specification	Quantity
Power supply/+12 V/1 A	1
Arduino Nano	1
Accelerometer	1
Accelerometer patch	1
LCD (20 * 4)	1
LCD patch	1
2.4 GHz RF modem	1
2.4 GHz RF modem patch	1
Connecting wires (M–M, M–F, F–F)	20 each
Zero-size PCB or bread board or a designed PCB	1

TABLE 16.2

Components List for a Receiver Section

Component/Specification	Quantity
Power supply/+12 V/1 A	1
Arduino Nano	1
Transistor 2N222	1
Relay	2
Bulb	1
Fan	1
LCD (20 * 4)	1
LCD patch	1
2.4 GHz RF modem	1
2.4 GHz RF modem patch	1
Connecting wires (M–M, M–F, F–F)	20 each
Zero-size PCB or bread board or a designed PCB	1

16.2 Circuit Diagram

To develop the system, connect the components as follows:

16.2.1 Transmitter Section

Arduino and Accelerometer
- Arduino GND-Module GND
- Arduino +5 V-Module +
- Arduino A0 pin-X pin of sensor
- Arduino A1 pin-Y pin of sensor

Arduino and RF modem
- Arduino pin0(TX)-RX pin of RF modem
- Arduino pin0(RX)-Tx pin of RF modem
- Arduino+5 V-+Vcc of RF modem
- Arduino Ground-GND of RF modem

Arduino and LCD
- Arduino digital pin 12-RS pin(4) of LCD
- Arduino digital pin GND-RW pin(5) of LCD
- Arduino digital pin 11-E pin(6) of LCD
- Arduino digital pin 10-D4 pin(11) of LCD
- Arduino digital pin 9-D5 pin(12) of LCD
- Arduino digital pin 8-D6 pin(13) of LCD
- Arduino digital pin 7-D7 pin(14) of LCD

16.2.2 Receiver Section

Arduino and RF modem
- Arduino pin0(TX)-RX pin of RF modem
- Arduino pin0(RX)-Tx pin of RF modem
- Arduino+5 V-+Vcc of RF modem
- Arduino Ground-GND of RF modem

Arduino and LCD
- Arduino digital pin 12-RS pin(4) of LCD
- Arduino digital pin GND-RW pin(5) of LCD
- Arduino digital pin 11-E pin(6) of LCD
- Arduino digital pin 10-D4 pin(11) of LCD
- Arduino digital pin 9-D5 pin(12) of LCD
- Arduino digital pin 8-D6 pin(13) of LCD
- Arduino digital pin 7-D7 pin(14) of LCD

2N2222, Relay and Arduino connection
- Arduino pin 6-base of 2N2222
- Collector of 2N2222-L2 end relay1
- L1 end of relay1-+12 V of battery or power supply
- COM pin of relay1-one end of AC LOAD
- NO pin of relay1-other end of AC LOAD
- Arduino pin 5-base of 2N2222
- Collector of 2N2222-L2 end relay2
- L1 end of relay2-+12 V of battery or power supply
- COM pin of relay2-one end of AC LOAD
- NO pin of relay2-other end of AC LOAD

Figure 16.2 shows the circuit diagram for the system.

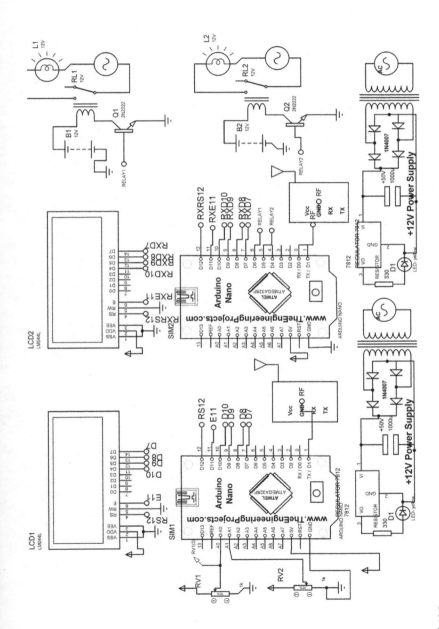

FIGURE 16.2
Circuit diagram.

16.3 Program

16.3.1 Transmitter Section

```
#include <LiquidCrystal.h>

LiquidCrystal lcd(12, 11, 10, 9, 8, 7);

int ACCELARATION_X_AXIS=A0;
int ACCELARATION_Y_AXIS=A1;
void setup()
{
Serial.begin(9600);// initialize serial communication
lcd.begin(20,4); // initialize LCD
lcd.setCursor(0,0);// choose Column and ROW of LCD
lcd.print("ACCELAROMETER BASED");//Print string on LCD
lcd.setCursor(0,1);// choose Column and ROW of LCD
lcd.print("AUTOMATION SYSTEM");// Print string on LCD
delay(1000);// delay of 1000mSec
}
void loop()
{
 int ACCELARATION_X_AXIS_LEVEL=analogRead(ACCELARATION_X_
AXIS);//read the level of sensor
 int ACCELARATION_Y_AXIS_LEVEL=analogRead(ACCELARATION_Y_
AXIS); //read the level of sensor
 int ACCELARATION_X_AXIS_BYTE=ACCELARATION_X_AXIS_LEVEL/10;
//scale the level by 10
 int ACCELARATION_Y_AXIS_BYTE=ACCELARATION_Y_AXIS_LEVEL/10;
//scale the level by 10

lcd.setCursor(0,2);// set cursor of LCD
lcd.print("XAXIS:");// print string on LCD
lcd.print(ACCELARATION_X_AXIS_BYTE);// print integer on LCD
lcd.setCursor(0,3);// set cursor of LCD
lcd.print("YAXIS:");// print string on LCD
lcd.print(ACCELARATION_Y_AXIS_BYTE);// print integer on LCD

Serial.print("XAXIS:");// print string on LCD
Serial.print(ACCELARATION_X_AXIS_BYTE);// print integer on LCD
Serial.print("YAXIS:");// print string on LCD
Serial.println(ACCELARATION_Y_AXIS_BYTE); print integer on LCD
delay(200);// delay 200mSec
}
```

16.3.2 Receiver Section

```
#include <LiquidCrystal.h>
iquidCrystal lcd(12, 11, 10, 9, 8, 7);
String inputString = "";          // choose string
int a=0;
int RELAY1=6;// define pin 6 is RELAY1
int RELAY2=5;// define pin 5 is RELAY2
void setup()
{
  Serial.begin(9600);// Initialize serial communication
  lcd.begin(20,4);// Initialise LCD
  pinMode(RELAY1,OUTPUT);// set pin 6 of Arduino as output
  pinMode(RELAY2,OUTPUT);// set pin 6 of Arduino as output
  lcd.setCursor(0,0);// set cursor
  lcd.print("ACCELAROMETER BASED");//print string on LCD
  lcd.setCursor(0,1);// set cursor of LCD
  lcd.print("AUTOMATION SYSTEM");// print string
  delay(1000);// delay 1000mSec
}

void loop()
{
 if (a==1)
 {
  Serial.println(inputString);// serial print
  lcd.setCursor(0,2);// set cursor
  lcd.print(inputString[0]);// print byte on LCD
  lcd.print(inputString[1]);// print byte on LCD
  lcd.print(inputString[2]);// print byte on LCD
  lcd.print(inputString[3]);// print byte on LCD
  lcd.print(inputString[4]);// print byte on LCD
  lcd.print(inputString[5]);// print byte on LCD
  lcd.print(inputString[6]);// print byte on LCD
  lcd.print(inputString[7]);// print byte on LCD
  lcd.setCursor(0,3);// set cursor
  lcd.print(inputString[8]);// print byte on LCD
  lcd.print(inputString[9]);// print byte on LCD
  lcd.print(inputString[10]);// print byte on LCD
  lcd.print(inputString[11]);// print byte on LCD
  lcd.print(inputString[12]);// print byte on LCD
  lcd.print(inputString[13]);// print byte on LCD
  lcd.print(inputString[14]);// print byte on LCD
  lcd.print(inputString[15]);// print byte on LCD
  if((inputString[6]>='4')&&(inputString[7]>='0'))
```

```
{
digitalWrite(RELAY1,HIGH);// set pin6 high
digitalWrite(RELAY2,LOW);//set pin6 high
delay(20);// delay 20mSec
}
if((inputString[14]>='5')&&(inputString[15]>='0'))
{
digitalWrite(RELAY2,HIGH);// set pin6 high
digitalWrite(RELAY1,LOW);// set pin6 low
delay(20);// delay 20mSec
}
if(inputString[0]==0x0A)
{
lcd.setCursor(0,2);// set cursor
lcd.print(inputString[1]);// print byte on LCD
lcd.print(inputString[2]); // print byte on LCD
lcd.print(inputString[3]); // print byte on LCD
lcd.print(inputString[4]); // print byte on LCD
lcd.print(inputString[5]); // print byte on LCD
lcd.print(inputString[6]); // print byte on LCD
lcd.print(inputString[7]); // print byte on LCD
lcd.print(inputString[8]); // print byte on LCD
lcd.setCursor(0,3);// set cursor
lcd.print(inputString[9]); // print byte on LCD
lcd.print(inputString[10]); // print byte on LCD
lcd.print(inputString[11]); // print byte on LCD
lcd.print(inputString[12]); // print byte on LCD
lcd.print(inputString[13]); // print byte on LCD
lcd.print(inputString[14]); // print byte on LCD
lcd.print(inputString[15]); // print byte on LCD
lcd.print(inputString[16]); // print byte on LCD
if((inputString[7]>='4')&&(inputString[8]>='0'))
{
digitalWrite(RELAY1,HIGH);// set pin 6 high
digitalWrite(RELAY2,LOW);//set pin 5 low
delay(20);
}
if((inputString[15]>='5')&&(inputString[16]>='0'))
{
digitalWrite(RELAY2,HIGH);// set pin 5 high
digitalWrite(RELAY1,LOW);// set pin 6 low
delay(20);// delay 20mSec
}
}
inputString = "";// clear the string
a=0;// set a variable 0
}

}
```

```
void serialEvent()
{
  while (Serial.available())// cheack serial
  {
   char inChar = (char)Serial.read(); // read serial
   inputString += inChar;// store byte in string
   if (inChar == 0x0D)
   {
     a=1;// set a=1
   }
  }
}
```

16.4 Proteus Simulation Model

Figure 16.3 shows the Proteus simulation model for interfacing the accelerometer with the Arduino. Load the program in Arduino and check the working of the circuit. The LCD displays the status of the system.

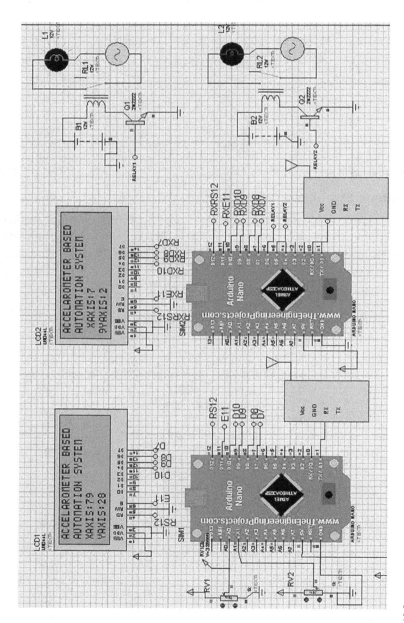

FIGURE 16.3
Proteus simulation model for the system.

17

Temperature Monitoring System Using RF Modem

17.1 Introduction

The project is about to design the temperature monitoring system using a RF modem. The system is designed in two sections: (1) transmitter section and (2) receiver section. Transmitter section comprises Arduino Uno, power supply, temperature sensor, LCD, and a RF modem. The receiver section comprises Arduino Uno, power supply, PC (LabVIEW), and RF modem. LabVIEW GUI is created to monitor the system. Figure 17.1 shows the block diagram of the system.

To design the system, following components are required (Tables 17.1 and 17.2).

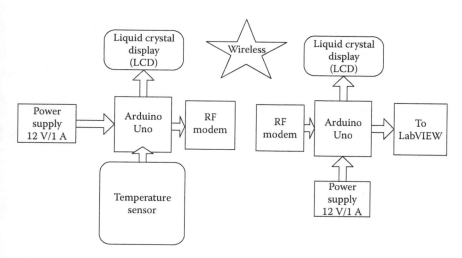

FIGURE 17.1
Block diagram of the system.

TABLE 17.1

Components List for a Transmitter Section

Component/Specification	Quantity
Power supply/+12 V/1 A	1
Arduino Uno	1
Temperature sensor	1
Temperature sensor patch	1
LCD (20 * 4)	1
LCD patch	1
2.4 GHz RF modem	1
2.4 GHz RF modem patch	1
Connecting wires (M–M, M–F, F–F)	20 each
Zero-size PCB or bread board or a designed PCB	1

TABLE 17.2

Components List for a Receiver Section

Component/Specification	Quantity
Power supply/+12 V/1 A	1
Arduino Uno	1
PC (LabVIEW)	1
LCD (20 * 4)	1
LCD patch	1
2.4 GHz RF modem	1
2.4 GHz RF modem patch	1
Connecting wires (M–M, M–F, F–F)	20 each
Zero-size PCB or bread board or a designed PCB	1

17.2 Circuit Diagram

To develop the system, connect the components as follows:

17.2.1 Transmitter Section

Arduino and LM35 sensor
- Arduino GND-Sensor GND

- Arduino +5 V-sensor +Vcc
- Arduino A0 pin-data out pin of sensor

Arduino and RF modem
- Arduino pin0(TX)-RX pin of RF modem
- Arduino pin0(RX)-Tx pin of RF modem
- Arduino+5 V-+Vcc of RF modem
- Arduino Ground-GND of RF modem

Arduino and LCD
- Arduino digital pin 12-RS pin(4) of LCD
- Arduino digital pin GND-RW pin(5) of LCD
- Arduino digital pin 11-E pin(6) of LCD
- Arduino digital pin 5-D4 pin(11) of LCD
- Arduino digital pin 4-D5 pin(12) of LCD
- Arduino digital pin 3-D6 pin(13) of LCD
- Arduino digital pin 2-D7 pin(14) of LCD

17.2.2 Receiver Section

Arduino and RF modem
- Arduino pin0(TX)-RX pin of RF modem
- Arduino pin0(RX)-Tx pin of RF modem
- Arduino+5 V-+Vcc of RF modem
- Arduino Ground-GND of RF modem

Arduino and LCD
- Arduino digital pin 12-RS pin(4) of LCD
- Arduino digital pin GND-RW pin(5) of LCD
- Arduino digital pin 11-E pin(6) of LCD
- Arduino digital pin 5-D4 pin(11) of LCD
- Arduino digital pin 4-D5 pin(12) of LCD
- Arduino digital pin 3-D6 pin(13) of LCD
- Arduino digital pin 2-D7 pin(14) of LCD

Figure 17.2 shows the circuit diagram for the system.

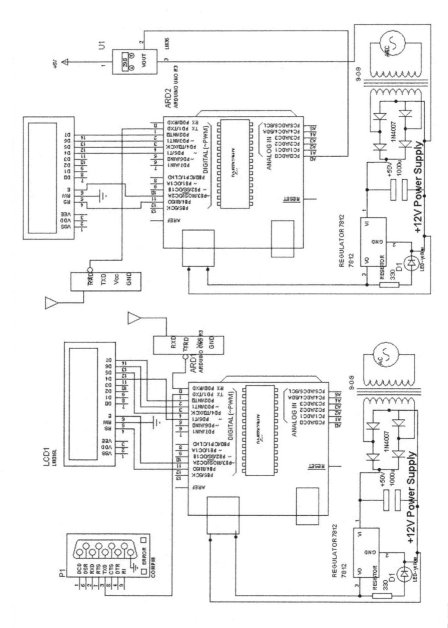

FIGURE 17.2

A circuit diagram.

17.3 Program

17.3.1 Transmitter Section

```
#include <LiquidCrystal.h>
LiquidCrystal lcd(12, 11, 5, 4, 3, 2);

void setup()
{
  lcd.begin(20,4);
  Serial.begin(9600);
  lcd.setCursor(0,0);
  lcd.print("Room temperature");
  lcd.setCursor(0,1);
  lcd.print("control System");
  delay(1000);
}

void loop()
{
  int TEMP_sensor_BYTE_LEVEL = analogRead(A0);
  int TEMP_sensor_BYTE=TEMP_sensor_BYTE_LEVEL/2;
  lcd.setCursor(0,2);
  lcd.print("Temperature 0C:");
  lcd.setCursor(0,3);
  lcd.print(TEMP_sensor_BYTE);
  Serial.write(TEMP_sensor_BYTE);

   delay(100);
}
```

17.3.2 Receiver Section

```
#include <LiquidCrystal.h>
LiquidCrystal lcd(13, 12, 11, 10, 9,8);

void setup()
{
  lcd.begin(20,4);
  Serial.begin(9600);
  lcd.setCursor(0,0);
  lcd.print("Room temperature");
  lcd.setCursor(0,1);
  lcd.print("control System");
  delay(1000);
}
```

```
// the loop routine runs over and over again forever:

void loop()
{
 int TEMP_sensor_BYTE=Serial.read();

 Serial.print("TEMPERATURESENSOR:");
 Serial.println(TEMP_sensor_BYTE);
   lcd.setCursor(0,2);
   lcd.print("Temperature 0C:");
   lcd.setCursor(0,3);
   lcd.print(TEMP_sensor_BYTE);
   delay(100);
}
```

17.4 Proteus Simulation Model

Figure 17.3 shows the Proteus simulation model for interfacing of a temperature sensor with the Arduino. Load the program in Arduino and check the working of the circuit. The LCD displays the status of the system.

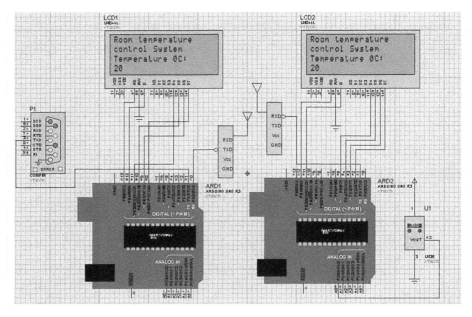

FIGURE 17.3
A Proteus simulation model.

17.5 LabVIEW GUI

LabVIEW GUI is designed for the temperature monitoring system. GUI is designed in two parts: (1) block diagram and (2) front panel. In the front panel, one thermometer is placed as the temperature indicator. It has serial COM port to access the data from the PC port, baud rate is set as "9600," which is same as the wireless modem to receive the data with the same rate, data bits "8," parity "none," stop bits "1," serial count is set as "100," and Serial_data_buffer to read the received data from the temperature sensor. Waveform chart is optional; it can be used to check the data that are received in the form of waveform. To make the front panel, an active graphical programming is done in the block diagram (Figures 17.4 through 17.6) with the help of blocks and steps discussed in Chapter 4.

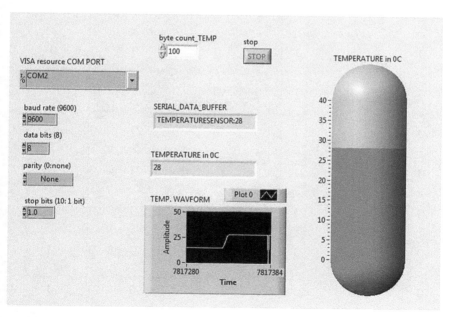

FIGURE 17.4

Front panel showing "Temperature 28°C."

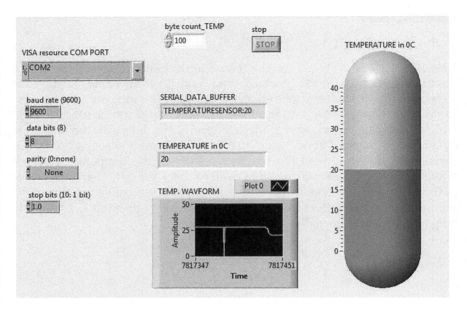

FIGURE 17.5
Front panel showing "Temperature 20°C."

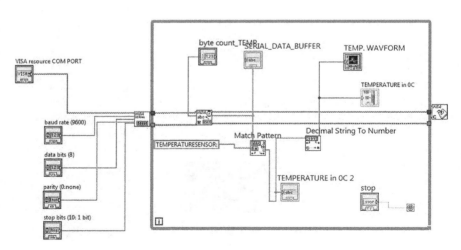

FIGURE 17.6
LabVIEW GUI block diagram for the system.

18

Emergency Hooter in the Case of a Disaster

18.1 Introduction

This project is about to generate an emergency alert in the case of a disaster. The system is designed in two sections: a sensor node and (2) a server. The sensor node comprises the Arduino Uno, power supply, soil moisture sensor, relay, hooter, LCD, and the RF modem. The relay is used to make the hooter "ON" or "OFF." The server comprises the Arduino Uno, power supply, switch array, PC (LabVIEW), and the RF modem. The switch array is attached to generate an emergency signal at the server and is sent to the sensor node. Figure 18.1 shows the block diagram of the system.

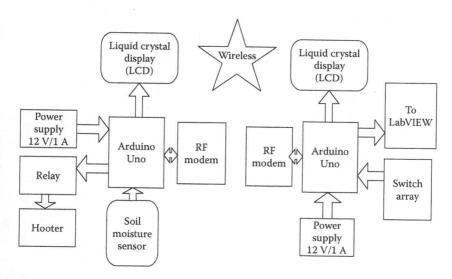

FIGURE 18.1
A block diagram of the system.

To design the system, following components are required (Tables 18.1 and 18.2).

TABLE 18.1

Components List for a Transmitter Section

Component/Specification	Quantity
Power supply/+12 V/1 A	1
Arduino Uno	1
Relay	1
Hooter	1
Soil moisture sensor	1
Soil moisture sensor patch	1
LCD (20 * 4)	1
LCD patch	1
2.4 GHz RF modem	1
2.4 GHz RF modem patch	1
Connecting wires (M–M, M–F, F–F)	20 each
Zero-size PCB or bread board or a designed PCB	1

TABLE 18.2

Components List for a Receiver Section

Component/Specification	Quantity
Power supply/+12 V/1 A	1
Arduino Uno	1
PC (LabVIEW)	1
Switch array	1
LCD (20 * 4)	1
LCD patch	1
2.4 GHz RF modem	1
2.4 GHz RF modem patch	1
Connecting wires (M–M, M–F, F–F)	20 each
Zero-size PCB or bread board or a designed PCB	1

18.2 Circuit Diagram

To develop the system, connect the components as follows:

18.2.1 Sensor Node

Arduino and SM sensor
- Arduino GND-Module GND
- Arduino +5 V-Module +
- Arduino A0 pin-data out pin of sensor

Arduino and Push button
- Arduino pin7-one terminal of 'MODE selection' push button
- Arduino pin6-one terminal of 'hooter ON' push button
- Arduino pin5-one terminal of 'hooter OFF' push button
- Other terminal of all push buttons-GND

Arduino and RF modem
- Arduino pin0(TX)-RX pin of RF modem
- Arduino pin0(RX)-Tx pin of RF modem
- Arduino+5 V-+Vcc of RF modem
- Arduino Ground-GND of RF modem

Arduino and LCD
- Arduino digital pin 13-RS pin(4) of LCD
- Arduino digital pin GND-RW pin(5) of LCD
- Arduino digital pin 12-E pin(6) of LCD
- Arduino digital pin 11-D4 pin(11) of LCD
- Arduino digital pin 10-D5 pin(12) of LCD
- Arduino digital pin 9-D6 pin(13) of LCD
- Arduino digital pin 8-D7 pin(14) of LCD

2N2222, relay, and Arduino connection

- Arduino pin 7-base of 2N2222
- Collector of 2N2222-L2 end relay
- L1 end of relay-+12 V of battery or power supply
- "COM" pin of relay-one end of AC Hooter
- "NO" pin of relay-other end of AC Hooter

18.2.2 Server

Arduino and RF modem

- Arduino pin0(TX)-RX pin of RF modem
- Arduino pin0(RX)-Tx pin of RF modem
- Arduino+5 V-+Vcc of RF modem
- Arduino Ground-GND of RF modem

Arduino and LCD

- Arduino digital pin 12-RS pin(4) of LCD
- Arduino digital pin GND-RW pin(5) of LCD
- Arduino digital pin 11-E pin(6) of LCD
- Arduino digital pin 5-D4 pin(11) of LCD
- Arduino digital pin 4-D5 pin(12) of LCD
- Arduino digital pin 3-D6 pin(13) of LCD
- Arduino digital pin 2-D7 pin(14) of LCD

Figure 18.2 shows the circuit diagram for the system.

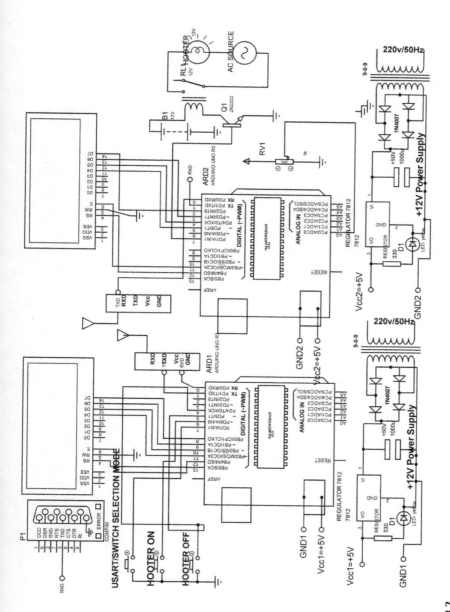

FIGURE 18.2
A circuit diagram.

18.3 Program

18.3.1 Sensor Node

```
#include <LiquidCrystal.h>
LiquidCrystal lcd(13, 12, 11, 10, 9, 8);
#define BUTTON_MODE_SELECTION 7
#define BUTTON_PIN_HOOTERON 6
#define BUTTON_PIN_HOOTEROFF 5
String inputString = "";     // a string to hold incoming data
boolean stringComplete = false;  // whether the string is
complete
void setup()
{
  lcd.begin(20,4);
  Serial.begin(9600);
  pinMode(BUTTON_MODE_SELECTION,INPUT_PULLUP);
  pinMode(BUTTON_PIN_HOOTERON,INPUT_PULLUP);
  pinMode(BUTTON_PIN_HOOTEROFF,INPUT_PULLUP);
  inputString.reserve(200);
  lcd.setCursor(0,0);
  lcd.print("Hooter control");
  lcd.setCursor(0,1);
  lcd.print("system in diste");
  delay(1000);
}

void loop()
{
  int BUTTON_MODE_SELECTION_digital=digitalRead
  (BUTTON_MODE_SELECTION);
  if (BUTTON_MODE_SELECTION_digital==LOW)
    {
      Serial.println(30);
        if (stringComplete)
        {
            lcd.clear();
            lcd.setCursor(0,3);
            lcd.print(inputString);
            delay(10);

            inputString = "";
            stringComplete = false;
        }
    }
```

```
else if(BUTTON_MODE_SELECTION_digital==HIGH)
    {
      int BUTTON_PIN_HOOTERON_digital= digitalRead
      (BUTTON_PIN_HOOTERON) ;
      int BUTTON_PIN_HOOTEROFF_digital= digitalRead
      (BUTTON_PIN_HOOTEROFF) ;
      if(BUTTON_PIN_HOOTERON_digital==LOW)
          {
          lcd.clear() ;
          lcd.setCursor(0,2) ;
          lcd.print("HOOTER ON") ;
          Serial.println(10) ;
          }

      if(BUTTON_PIN_HOOTEROFF_digital==LOW)
          {
          lcd.clear() ;
          lcd.setCursor(0,2) ;
          lcd.print("HOOTER OFF") ;
          Serial.println(20) ;
          }
    }
}
void serialEvent()
    {
      while (Serial.available())
      {
        char inChar = (char)Serial.read() ;
        inputString += inChar;
      if(inChar == 0x0D)
          {
          stringComplete = true;
          }
      }
    }
```

18.3.2 Server

```
#include <LiquidCrystal.h>
LiquidCrystal lcd(12, 11, 5, 4, 3, 2);
#define HOOTER_RELAY 7
#define SM_SENSOR A0
String inputString = "";        // a string to hold
incoming data
boolean stringComplete = false;  // whether the string
is complete
```

```
void setup()
{
  Serial.begin(9600);
  lcd.begin(20,4);
  pinMode(HOOTER_RELAY, OUTPUT);
  inputString.reserve(200);
  lcd.setCursor(0,0);
  lcd.print("Hooter control");
  lcd.setCursor(0,1);
  lcd.print("system in dister");
  delay(1000);
}

void loop()
{
  if(stringComplete)
      {
          lcd.clear();
          delay(10);
              if((inputString[0]=='1')&&(inputString[1]=='0')
                {
                  lcd.clear();
                  lcd.setCursor(0,2);
                  lcd.print("HOOTER ON");
                  digitalWrite(HOOTER_RELAY, HIGH);
                  delay(20);
                }
                else if ((inputString[0]=='2')&&
                (inputString[1]=='0'))
                {
                  lcd.clear();
                  lcd.setCursor(0,2);
                  lcd.print("HOOTER OFF");
                  digitalWrite(HOOTER_RELAY, LOW);
                  delay(20);

                }
                else if ((inputString[0]=='3')&&
                (inputString[1]=='0'))
                {
                  int ANALOG_READ_SM_SENSOR_ LEVEL=analogRead
                  (SM_SENSOR);
                  int ANALOG_READ_SM_SENSOR_ PPM=ANALOG_READ_
                  SM_SENSOR_LEVEL/10;
                  lcd.setCursor(0,3);
                  lcd.print("SMSENSOR:");
                  lcd.print(ANALOG_READ_SM_ SENSOR_PPM);
                  Serial.print("SMSENSOR:");
                  Serial.println(ANALOG_READ_ SM_SENSOR_PPM);
```

```
        delay(20);
      }
    if((inputString[0]==0x0A))
      {
        if ((inputString[1]=='1')&& (inputString[2]
        =='0'))

        {
          lcd.clear();
          lcd.setCursor(0,2);
          lcd.print("HOOTER ON");
          digitalWrite(HOOTER_RELAY, HIGH);
          delay(20);
        }

        else if ((inputString[1]=='2')&&(inputString[2]
        =='0'))
        {
          lcd.clear();
          lcd.setCursor(0,2);
          lcd.print("HOOTER OFF");
          digitalWrite(HOOTER_RELAY, LOW);
          delay(20);

        }
        else if ((inputString[1]=='3') &&
        (inputString[2]=='0'))
        {
          int ANALOG_READ_SM_SENSOR_LEVEL=analogRead
          (SM_SENSOR);
          int ANALOG_READ_SM_SENSOR_PPM=ANALOG_READ_
          SM_SENSOR_ LEVEL/10;
          lcd.setCursor(0,3);
          lcd.print("SMSENSOR:");
          lcd.print(ANALOG_READ_SM_SENSOR_PPM);
          Serial.print("SMSENSOR:");
          Serial.println(ANALOG_READ_SM_SENSOR_PPM);
          delay(20);
        }


      }



inputString = "";
stringComplete = false;
}
```

```
delay(10);
}

void serialEvent()
{
  while (Serial.available())
  {
    char inChar = (char)Serial.read();
    inputString += inChar;
  if (inChar == 0x0D)
    {
    stringComplete = true;
    }
  }
}
```

18.4 Proteus Simulation Model

Figure 18.3 shows the Proteus simulation model for interfacing the soil moisture sensor with the Arduino. Load the program in the Arduino and check the working of the circuit. A LCD displays the status of the system. In Figure 18.4, a Proteus simulation model shows the hooter "ON." In Figure 18.5, a Proteus simulation model shows the hooter "OFF."

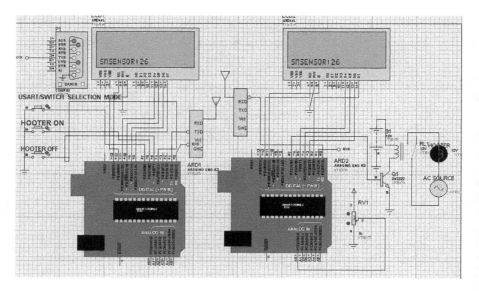

FIGURE 18.3
A Proteus simulation model showing the soil moisture content.

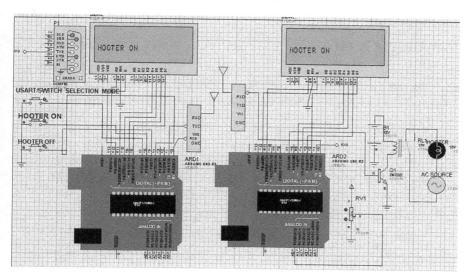

FIGURE 18.4
A Proteus simulation model showing hooter "ON."

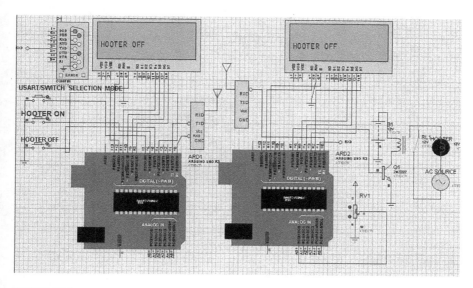

FIGURE 18.5
A Proteus simulation model showing hooter "OFF."

18.5 LabVIEW GUI

LabVIEW GUI is designed for the disaster monitoring system. GUI is designed in two parts: (1) block diagram and (2) front panel. The front panel has a serial COM port to access the data from the PC port, the baud rate is set as "9600," which is same as the wireless modem "Zigbee" to receive the data with the same rate, data bits "8," parity "none," stop bits "1," serial count is set as "100," flow control "None," and Serial_read_buffer to read the received data from the fire sensor. Waveform chart is optional; it can be used to check the data received in the form of waveform. To make the front panel, an active graphical programming is done in the block diagram (Figures 18.6 and 18.7) with the help of blocks and steps discussed in Chapter 4.

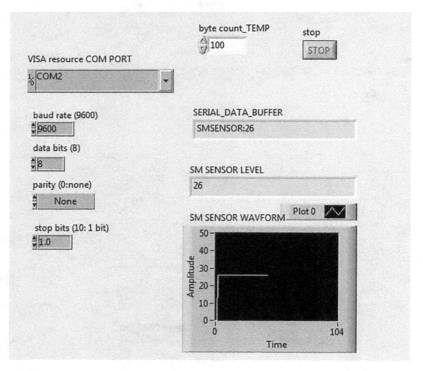

FIGURE 18.6
Front panel for the LabVIEW GUI.

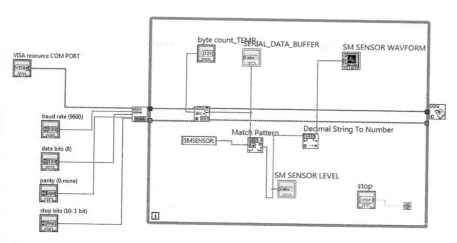

FIGURE 18.7
LabVIEW GUI block diagram for the system.

19

LabVIEW GUI-Based Wireless Robot Control System

19.1 Introduction

This chapter explains a project titled "LabVIEW GUI-based robot control system." The system is designed in two sections: (1) transmitter section and (2) receiver section. The transmitter section comprises the Arduino Uno, power supply, PC (LabVIEW), LCD, and the RF modem. The receiver section comprises the Arduino Uno, power supply, LCD, DC motor, and RF modem. In this project, two independent methods to control a DC motor are discussed. A motor can be controlled with a switch array as well as with LabVIEW GUI for "clockwise," "anticlockwise," and "stop" to make movement in the forward or reverse direction. The receiver section comprises the Arduino Uno, power supply, motor driver (L293D), DC motor, and RF modem. A chassis is used to make the robot body. A free wheel is connected in the front side of the robot with two DC motor at the back sides. Figure 19.1 shows the block diagram of the system.

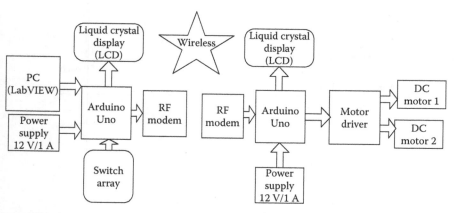

FIGURE 19.1
Block diagram of the Robot.

To design the system, following components are required (Tables 19.1 and 19.2).

TABLE 19.1

Components List for a Transmitter Section

Component/Specification	Quantity
Power supply/+12 V/1 A	1
PC (LabVIEW)	1
Arduino Uno	1
LCD (20 * 4)	1
LCD patch	1
2.4 GHz RF modem	1
2.4 GHz RF modem patch	1
Connecting wires (M–M, M–F, F–F)	20 each
Zero-size PCB or bread board or a designed PCB	1

TABLE 19.2

Components List for a Receiver Section

Component/Specification	Quantity
Power supply/+12 V/1 A	1
Arduino Uno	1
DC motor	2
L293D	1
LCD (20 * 4)	1
LCD patch	1
2.4 GHz RF modem	1
2.4 GHz RF modem patch	1
Connecting wires (M–M, M–F, F–F)	20 each
Zero-size PCB or bread board or a designed PCB	1

19.2 Circuit Diagram

To develop the system, connect the components as follows:

19.2.1 Transmitter Section

Arduino and RF modem
- Arduino pin0(TX)-RX pin of RF modem
- Arduino pin0(RX)-Tx pin of RF modem
- Arduino+5 V-+Vcc of RF modem
- Arduino Ground-GND of RF modem

Arduino and LCD
- Arduino digital pin 12-RS pin(4) of LCD
- Arduino digital pin GND-RW pin(5) of LCD
- Arduino digital pin 11-E pin(6) of LCD
- Arduino digital pin 5-D4 pin(11) of LCD
- Arduino digital pin 4-D5 pin(12) of LCD
- Arduino digital pin 3-D6 pin(13) of LCD
- Arduino digital pin 2-D7 pin(14) of LCD

19.2.2 Receiver Section

Arduino and RF modem
- Arduino pin0(TX)-RX pin of RF modem
- Arduino pin0(RX)-Tx pin of RF modem
- Arduino+5 V-+Vcc of RF modem
- Arduino Ground-GND of RF modem

Arduino and LCD
- Arduino digital pin 12-RS pin(4) of LCD
- Arduino digital pin GND-RW pin(5) of LCD
- Arduino digital pin 11-E pin(6) of LCD
- Arduino digital pin 5-D4 pin(11) of LCD

- Arduino digital pin 4-D5 pin(12) of LCD
- Arduino digital pin 3-D6 pin(13) of LCD
- Arduino digital pin 2-D7 pin(14) of LCD

Arduino and L293D connection
- Arduino GND-4,5,12,13 pins of IC
- Arduino +5 V-1,9,16 pins of IC
- Arduino pin 9-pin 2 of IC
- Arduino pin 8-pin 7 of IC
- Arduino pin 7-pin 10 of IC
- Arduino pin 6-pin 15 of IC
- L293D pin8-+ve of 12 V battery

L293D and DC motor connection
- L293D pin 3-to +ve pin of DC motor1
- L293D pin 6-to −ve pin of DC motor1
- L293D pin 10-to +ve pin of DC motor2
- L293D pin 15-to −ve pin of DC motor2

Figure 19.2 shows the circuit diagram for the system, connecting all the components as per connections.

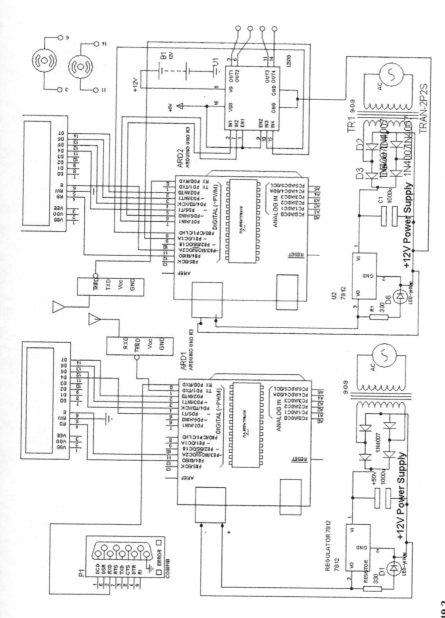

FIGURE 19.2

A circuit diagram.

19.3 Program

19.3.1 Transmitter Program

```
#include <LiquidCrystal.h>
LiquidCrystal lcd(13, 12, 11, 10, 9, 8);

void setup()
{
  lcd.begin(20,4);
  Serial.begin(9600);
  lcd.setCursor(0,0);
  lcd.print("DC motor control");
  lcd.setCursor(0,1);
  lcd.print("+ remote control");
  delay(1000);
}

void loop()
{
char LABVIEW_SERIAL_CHAR;
LABVIEW_SERIAL_CHAR=Serial.read();
  if(LABVIEW_SERIAL_CHAR=='W')

  {
    lcd.clear();
    lcd.setCursor(0,3);
    lcd.print("Motor Forward");
    Serial.write('W');

  }
  else if (LABVIEW_SERIAL_CHAR=='Z')
  {
    lcd.clear();
    lcd.setCursor(0,3);
    lcd.print("Motor Reverse");
    Serial.write('Z');

  }
  else if (LABVIEW_SERIAL_CHAR=='X')
  {
    lcd.clear();
    lcd.setCursor(0,3);
    lcd.print("Motor Left");
    Serial.write('X');

  }
  else if (LABVIEW_SERIAL_CHAR=='Y')
```

```
  {
   lcd.clear();
   lcd.setCursor(0,3);
   lcd.print("Motor Right");
   Serial.write('Y');

  }
  else if (LABVIEW_SERIAL_CHAR=='E')
  {
   lcd.clear();
   lcd.setCursor(0,3);
   lcd.print("STOP");
   Serial.write('E');
  }
   delay(10);
 }
```

19.3.2 Receiver Program

```
#include <LiquidCrystal.h>
LiquidCrystal lcd(12, 11, 5, 4, 3, 2);
#define DC_MOTOR1_POSITIVE 6
#define DC_MOTOR1_NEGATIVE 7
#define DC_MOTOR2_POSITIVE 8
#define DC_MOTOR2_NEGATIVE 9
void setup()
{
  Serial.begin(9600);
  lcd.begin(20,4);
  pinMode(DC_MOTOR1_POSITIVE, OUTPUT);
  pinMode(DC_MOTOR1_NEGATIVE, OUTPUT);
  pinMode(DC_MOTOR2_POSITIVE, OUTPUT);
  pinMode(DC_MOTOR2_NEGATIVE, OUTPUT);
  lcd.setCursor(0,0);
  lcd.print("DC motor control");
  lcd.setCursor(0,1);
  lcd.print("+ remote control");
  delay(1000);
}

void loop()
{
char LABVIEW_SERIAL_CHAR;
LABVIEW_SERIAL_CHAR=Serial.read();
  if(LABVIEW_SERIAL_CHAR=='W')

  {
    lcd.clear();
    lcd.setCursor(0,3);
```

```
  lcd.print("Motor Forward");
  digitalWrite(DC_MOTOR1_POSITIVE, HIGH);
  digitalWrite(DC_MOTOR1_NEGATIVE, LOW);
  digitalWrite(DC_MOTOR2_POSITIVE, HIGH);
  digitalWrite(DC_MOTOR2_NEGATIVE, LOW);
}
else if (LABVIEW_SERIAL_CHAR=='Z')
{
  lcd.clear();
  lcd.setCursor(0,3);
  lcd.print("Motor Reverse");
  digitalWrite(DC_MOTOR1_POSITIVE, LOW);
  digitalWrite(DC_MOTOR1_NEGATIVE, HIGH);
  digitalWrite(DC_MOTOR2_POSITIVE, LOW);
  digitalWrite(DC_MOTOR2_NEGATIVE, HIGH);
}
else if (LABVIEW_SERIAL_CHAR=='X')
{
  lcd.clear();
  lcd.setCursor(0,3);
  lcd.print("Motor Left");
  digitalWrite(DC_MOTOR1_POSITIVE, HIGH);
  digitalWrite(DC_MOTOR1_NEGATIVE, LOW);
  digitalWrite(DC_MOTOR2_POSITIVE, LOW);
  digitalWrite(DC_MOTOR2_NEGATIVE, LOW);
}
else if (LABVIEW_SERIAL_CHAR=='Y')
{
  lcd.clear();
  lcd.setCursor(0,3);
  lcd.print("Motor Right");
  digitalWrite(DC_MOTOR1_POSITIVE, LOW);
  digitalWrite(DC_MOTOR1_NEGATIVE, LOW);
  digitalWrite(DC_MOTOR2_POSITIVE, HIGH);
  digitalWrite(DC_MOTOR2_NEGATIVE, LOW);
}

else if (LABVIEW_SERIAL_CHAR=='E')
{
  lcd.clear();
  lcd.setCursor(0,3);
  lcd.print("STOP");
  digitalWrite(DC_MOTOR2_POSITIVE, LOW);
  digitalWrite(DC_MOTOR1_NEGATIVE, LOW);
  digitalWrite(DC_MOTOR2_POSITIVE, LOW);
  digitalWrite(DC_MOTOR2_NEGATIVE, LOW);
}

  delay(10);
}
```

19.4 Proteus Simulation Model

Figure 19.3 shows the Proteus simulation model for the system. Load the program in the Arduino and check the working of the circuit. The LCD displays the status of the system.

19.5 LabVIEW GUI

LabVIEW GUI is designed to control the Robot. GUI is designed in two parts: (1) block diagram and the (2) front panel. In the front panel, five buttons are used to make the robot moving in "forward," "reverse," "left," "right," and "stop." It has a serial COM port to access the data from the PC port, baud rate is set as "9600," which is same as the wireless modem "Zigbee" to send the data with the same rate, data bits "8," parity "none," and stop bits "1." To make the front panel, an active graphical programming is done in the block diagram (Figures 19.4 and 19.5) with the help of blocks and steps discussed in Chapter 4.

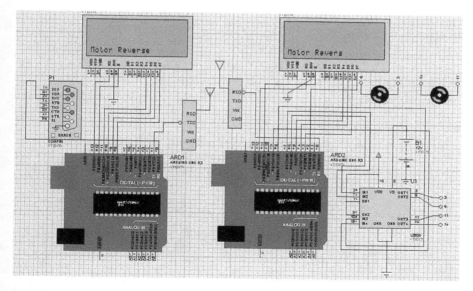

FIGURE 19.3
A Proteus simulation model showing the motor in reverse direction.

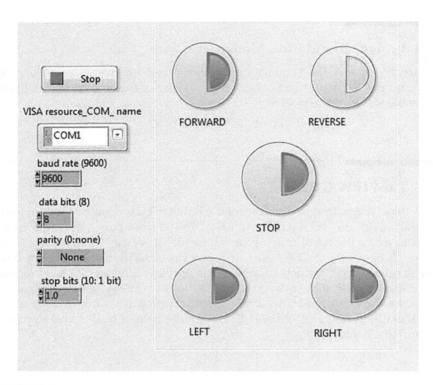

FIGURE 19.4
Front panel to control the Robot.

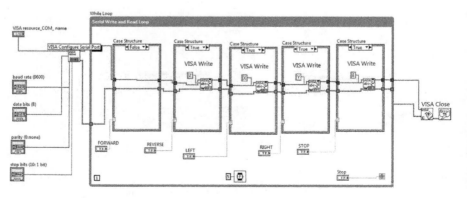

FIGURE 19.5
LabVIEW GUI block diagram for the system.

20

Home Automation System Using DTMF

20.1 Introduction

This project is designed to control the home appliances with dual-tone multi-frequency decoder (DTMF). The system has two sections: (1) receiver section and (2) remote control (mobile). The system is designed to control (ON & OFF) four loads through relays. The receiver section comprises Arduino Nano, power supply, LCD, DTMF decoder, mobile phone, and the head phone with a 3 mm jack. Another mobile phone is used as a remote control. Figure 20.1 shows the block diagram of the system.

To design the system, following components are required (Table 20.1).

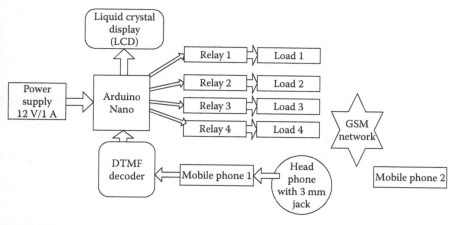

FIGURE 20.1
Block diagram of the system.

TABLE 20.1

Components List for a Receiver Section

Component/Specification	Quantity
Power supply/+12 V/1 A	1
Arduino Nano	1
Relay 12 V/1 A	4
Transistor 2N2222	4
DTMF decoder	1
3 mm audio jack	1
LCD (20 * 4)	1
LCD patch	1
AC plug	4
Bulb	4
Bulb holder	4
Connecting wires (M–M, M–F, F–F)	20 each
Zero-size PCB or bread board or a designed PCB	1

20.2 Circuit Diagram

To develop the system, connect the components as follows:

Arduino and Push button

- Arduino pin7-connect message button
- Arduino pin6-connect CALL button
- Arduino pin5-connect END button

Arduino and DTMF Decoder

- Arduino pin6-LED1 pin of decoder
- Arduino pin5-LED2 pin of decoder
- Arduino pin4-LED3 pin of decoder
- Arduino pin3-LED4 pin of decoder
- Arduino+5 V-+Vcc of decoder
- Arduino Ground-GND of decoder

2N2222, Relay and Arduino connection

- Arduino pin 2-base of 2N2222
- Collector of 2N2222-L2 end relay1
- "L1" end of relay1 to +12 V of battery or power supply
- "COM" pin of relay1-one end of AC load
- "NO" pin of relay1-other end of AC load
- Arduino pin 1-base of 2N2222
- Collector of 2N2222-"L2" end relay2
- "L1" end of relay2-+12 V of battery or power supply
- "COM" pin of relay2-one end of AC load
- "NO" pin of relay2-other end of AC load
- Arduino pin 0-base of 2N2222
- Collector of 2N2222-"L2" end relay3
- "L1" end of relay3-+12 V of battery or power supply
- "COM" pin of relay3-one end of AC load
- "NO" pin of relay3-other end of AC load
- Arduino pin 13-base of 2N2222
- Collector of 2N2222-"L2" end relay4
- "L1" end of relay4-+12 V of battery or power supply
- "COM" pin of relay4-one end of AC load
- "NO" pin of relay4-other end of AC load

Arduino and LCD

- Arduino digital pin 12-RS pin(4) of LCD
- Arduino digital pin GND-RW pin(5) of LCD
- Arduino digital pin 11-E pin(6) of LCD
- Arduino digital pin 10-D4 pin(11) of LCD
- Arduino digital pin 9-D5 pin(12) of LCD
- Arduino digital pin 8-D6 pin(13) of LCD
- Arduino digital pin 7-D7 pin(14) of LCD

Figure 20.2 shows the circuit diagram for the system.

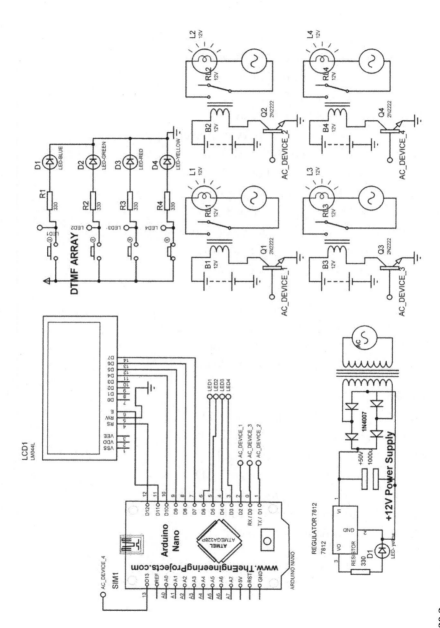

FIGURE 20.2
Circuit diagram for the system.

20.3 Program

```
// include the library code:
#include <LiquidCrystal.h>
LiquidCrystal lcd(12, 11, 10, 9, 8, 7);// initialize the LCD
Library w.r t. RS,E,D4,D5,D6,D7
int DTMF_LED1_HIGH=6;
int DTMF_LED2_HIGH=5;
int DTMF_LED3_HIGH=4;
int DTMF_LED4_HIGH=3;
int DTMF_LED1_HIGH_READ=0;
int DTMF_LED2_HIGH_READ=0;
int DTMF_LED3_HIGH_READ=0;
int DTMF_LED4_HIGH_READ=0;
int AC_DEVICE_1=2;
int AC_DEVICE_2=1;
int AC_DEVICE_3=0;
int AC_DEVICE_4=13;
void setup()
{
 pinMode(DTMF_LED1_HIGH, INPUT);//configure pin6 as an input
 and enable the internal pull-up resistor
 pinMode(DTMF_LED2_HIGH, INPUT);//configure pin6 as an input
 and enable the internal pull-up resistor
 pinMode(DTMF_LED3_HIGH, INPUT);//configure pin6 as an input
 and enable the internal pull-up resistor
 pinMode(DTMF_LED4_HIGH, INPUT);//configure pin6 as an input
 and enable the internal pull-up resistor
 pinMode(AC_DEVICE_1,OUTPUT);
 pinMode(AC_DEVICE_2,OUTPUT);
 pinMode(AC_DEVICE_3,OUTPUT);
 pinMode(AC_DEVICE_4,OUTPUT);
 lcd.begin(20, 4);// set up the LCD's number of columns and
 rows
 lcd.setCursor(0, 0);// set cursor to column0 and row1
 lcd.print("DTMF BASED HOME ");// Print a message to the LCD.
 lcd.setCursor(0, 1);// set cursor to column0 and row1
 lcd.print("AUTOMATION SYSTEM......");// Print a message to
 the LCD.
 delay(1000);
}
```

```
void loop()
{

 DTMF_LED1_HIGH_READ = digitalRead(DTMF_LED1_HIGH);//read the
 pushbutton value into a variable
 DTMF_LED2_HIGH_READ = digitalRead(DTMF_LED2_HIGH);//read the
 pushbutton value into a variable
 DTMF_LED3_HIGH_READ = digitalRead(DTMF_LED3_HIGH);//read the
 pushbutton value into a variable
 DTMF_LED4_HIGH_READ = digitalRead(DTMF_LED4_HIGH);//read the
 pushbutton value into a variable

 if ((DTMF_LED1_HIGH_READ == LOW)&&(DTMF_LED2_HIGH_READ ==
 LOW)&&(DTMF_LED3_HIGH_READ == LOW)&&(DTMF_LED4_HIGH_READ ==
 HIGH))// Read PIN 6,5,4,3 as HIGH PIN
 {
  lcd.clear();
  lcd.setCursor(0, 2);// set cursor to column0 and row2
  lcd.print("NUMBER 1 FROM MOBILE");// Print a message to
  the LCD.
  digitalWrite(AC_DEVICE_1,HIGH);
  digitalWrite(AC_DEVICE_2,LOW);
  digitalWrite(AC_DEVICE_3,LOW);
  digitalWrite(AC_DEVICE_4,LOW);
  lcd.setCursor(0, 3);
  lcd.print("AC_DEVICE_1 ON");
  delay(20);
 }
 if ((DTMF_LED1_HIGH_READ == LOW)&&(DTMF_LED2_HIGH_READ ==
 LOW)&&(DTMF_LED3_HIGH_READ ==HIGH)&&(DTMF_LED4_HIGH_READ ==
 LOW))// Read PIN 6,5,4,3 as HIGH PIN
 {
  lcd.clear();
  lcd.setCursor(0, 2);// set cursor to column0 and row2
  lcd.print("NUMBER 2 FROM MOBILE");// Print a message to
  the LCD.
  digitalWrite(AC_DEVICE_1,LOW);
  digitalWrite(AC_DEVICE_2,HIGH);
  digitalWrite(AC_DEVICE_3,LOW);
  digitalWrite(AC_DEVICE_4,LOW);
  lcd.setCursor(0, 3);
  lcd.print("AC_DEVICE_2 ON");
  delay(20);
 }


 if ((DTMF_LED1_HIGH_READ == LOW)&&(DTMF_LED2_HIGH_READ ==
 LOW)&&(DTMF_LED3_HIGH_READ ==HIGH)&&(DTMF_LED4_HIGH_READ ==
 HIGH))// Read PIN 6,5,4,3 as HIGH PIN
```

```
  {
   lcd.clear();
   lcd.setCursor(0, 2);// set cursor to column0 and row2
   lcd.print("NUMBER 3 FROM MOBILE");// Print a message to the
   LCD.
   digitalWrite(AC_DEVICE_1,LOW);
   digitalWrite(AC_DEVICE_2,LOW);
   digitalWrite(AC_DEVICE_3,HIGH);
   digitalWrite(AC_DEVICE_4,LOW);
   lcd.setCursor(0, 3);
   lcd.print("AC_DEVICE_3 ON");
   delay(20);
  }
  if ((DTMF_LED1_HIGH_READ == LOW)&&(DTMF_LED2_HIGH_READ ==
  HIGH)&&(DTMF_LED3_HIGH_READ ==LOW)&&(DTMF_LED4_HIGH_READ ==
  LOW))// Read PIN 6,5,4,3 as HIGH PIN
  {
   lcd.clear();
   lcd.setCursor(0, 2);// set cursor to column0 and row2
   lcd.print("NUMBER 4 FROM MOBILE");// Print a message to
   the LCD.
   digitalWrite(AC_DEVICE_1,LOW);
   digitalWrite(AC_DEVICE_2,LOW);
   digitalWrite(AC_DEVICE_3,LOW);
   digitalWrite(AC_DEVICE_4,HIGH);
   lcd.setCursor(0, 3);
   lcd.print("AC_DEVICE_4 ON");
   delay(20);
  }

  if ((DTMF_LED1_HIGH_READ == LOW)&&(DTMF_LED2_HIGH_READ ==
  HIGH)&&(DTMF_LED3_HIGH_READ ==LOW)&&(DTMF_LED4_HIGH_READ ==
  HIGH))// Read PIN 6,5,4,3 as HIGH PIN
  {
   lcd.clear();
   lcd.setCursor(0, 2);// set cursor to column0 and row2
   lcd.print("NUMBER 5 FROM MOBILE");// Print a message to the
   LCD.
   digitalWrite(AC_DEVICE_1,LOW);
   digitalWrite(AC_DEVICE_2,LOW);
   digitalWrite(AC_DEVICE_3,LOW);
   digitalWrite(AC_DEVICE_4,LOW);
   lcd.setCursor(0, 3);
   lcd.print("AC_DEVICE OFF");
   delay(20);
  }
}
```

20.4 Proteus Simulation Model

Figure 20.3 shows the Proteus simulation model for interfacing the DTMF decoder with the Arduino. Load the program in the Arduino and check the working of the circuit. The LCD displays the status of the system.

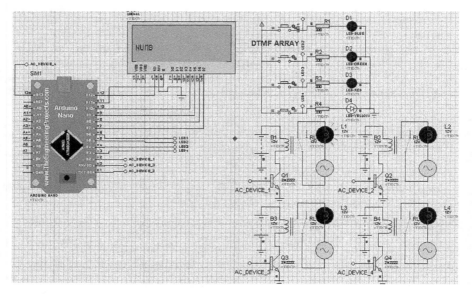

FIGURE 20.3
Proteus simulation model of the receiver section.

21

RFID Card-Based Attendance System

21.1 Introduction

The objective of this chapter is to develop a RFID card-based attendance system. Each student will be having a RFID tag with a unique ID, which will be first identified and then stored in the program. The stored id will be matched with the RFID tag ID, which will be swiped to the reader to authenticate the presence of a student. The system comprises Arduino Uno, power supply, RFID reader, and the RFID tag. It operates on a baud rate of 9600. It can be connected to an RS232, which is compatible to the PC or to the UART. RFID tag generates a 12 byte unique ID. Figure 21.1 shows the block diagram of the system.

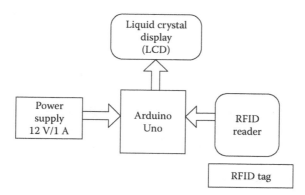

FIGURE 21.1
Block diagram of the system.

To design the system, following components are required (Table 21.1).

TABLE 21.1

Components List for a System

Component/Specification	Quantity
Power supply/+12 V/1 A	1
Arduino Uno	1
RFID reader	1
RFID tag	1
MAX232 board (if RFID reader is with max232)	1
LCD (20 * 4)	1
LCD patch	1
2.4 GHz RF modem	1
2.4 GHz RF modem patch	1
Connecting wires (M–M, M–F, F–F)	20 each
Zero-size PCB or bread board or a designed PCB	1

21.2 Circuit Diagram

21.2.1 Steps to Read RFID Reader

- Connect female DB9 connector to the RFID reader as shown in Figure 21.2.
- Connect the serial cable to female DB9. Connect the other end to the Com Port of the PC.
- Connect the power Supply (7-9V DC). Then the LED on the reader (PWR) will be 'ON' and the LED on the reader (STS) will blink.
- The Baud rate is set as 9600.
- The extracted code can be checked on serial monitor of Arduino IDE.
- Use this extracted code in the program.
- More than one RFID card code can be written in a single program.

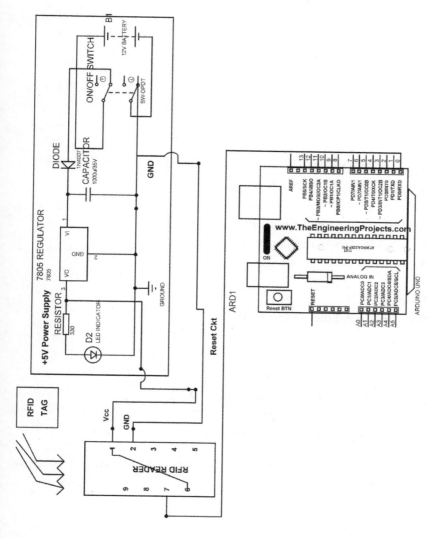

FIGURE 21.2
Connections to read the RFID reader.

21.2.1.1 Program to Extract the Code

```
void setup()
{
Serial.begin(9600);
}
void loop()
{
int BYTE_RFID = Serial.read();
Serial.print(BYTE_RFID );
}
```

21.2.2 Connections

To develop the system, connect the components as follows:

Arduino and RFID reader
- Arduino GND-GND(2) of Module
- Arduino +5 V-+Vcc (1) of Module
- Arduino Tx Pin-Rx pin of module
- Arduino Rx Pin-Tx out pin(7) of module
- Arduino +5 V-SEL pin(6) of module

Arduino and LCD
- Arduino digital pin 12-RS pin(4) of LCD
- Arduino digital pin GND-RW pin(5) of LCD
- Arduino digital pin 11-E pin(6) of LCD
- Arduino digital pin 10-D4 pin(11) of LCD
- Arduino digital pin 9-D5 pin(12) of LCD
- Arduino digital pin 8-D6 pin(13) of LCD
- Arduino digital pin 7-D7 pin(14) of LCD

Figure 21.3 shows the circuit diagram for the system, connecting all the components as per connections.

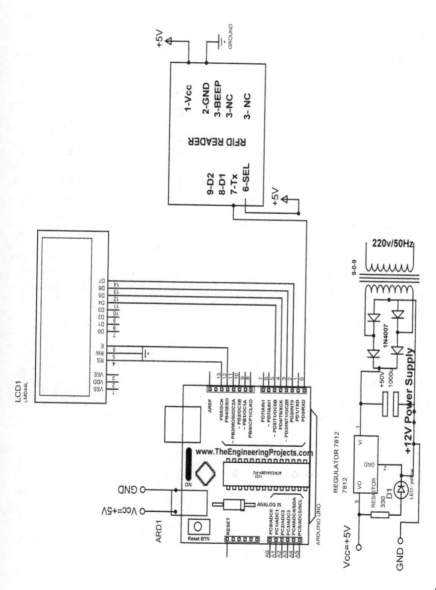

FIGURE 21.3
Circuit Diagram to interface RFID with Arduino.

21.3 Program

```
#include<SoftwareSerial.h>
#include <LiquidCrystal.h>
SoftwareSerial mySerial(9,10);
LiquidCrystal lcd(13, 12, 5, 4, 3, 2);
int j=0,k=0;
int read_count=0;
char data_temp, RFID_data[12],x=0;
char Saved_Tags[3][12]={
        {'5','0','0','0','9','2','B','E','9','3','E','F'},
        //ASCII value of extracted code from RFID card1
        {'5','0','0','0','9','2','E','A','2','C','0','4'},
        //ASCII value of extracted code from RFID card2
        {'5','0','0','0','9','3','2','1','7','F','9','D'}
        //ASCII value of extracted code from RFID card3
          };

boolean tag_check,tag_status,entry_control;
void setup()
{
  lcd.begin(20,4);
  mySerial.begin(9600);
  Serial.begin(9600);
  lcd.clear();
  lcd.setCursor(0,0);
  lcd.print("RFID CARD READER");
  lcd.setCursor(0,1);
  lcd.print("and data logger");
  delay(1000);
  lcd.clear();
  lcd.setCursor(0,0);
  lcd.print("PUNCH THE CARD ");
}

void loop()
{
  RecieveData();
  CheckData();
  AccessCheck();
}

void RecieveData()
{
 if(mySerial.available()>0)
   {
   data_temp=mySerial.read();
   RFID_data[read_count]=data_temp;
```

```
      read_count++;
      lcd.setCursor(0,2);
      lcd.print(RFID_data[read_count]);
      Serial.print(RFID_data[read_count]);
     }
  }

 void CheckData()
  {
  if(read_count==12)
   {
   entry_control=true;
   for(k=0;k<3;k++)
   {
   for(j=0;j<12;j++)
    {
    if(Saved_Tags[k][j]==RFID_data[j])
     {
     tag_check=true;
     }
     else
      {
      tag_check=false;
      break;
      }
    }
  if(tag_check==true)
    {
   tag_status=true;
    }
   }
  read_count=0;
   }
  }

void AccessCheck()
{
 if(entry_control==true)
 {
  if(tag_status==true)
  {
   x=1;
   lcd.setCursor(0,1);
   lcd.print("ACCESS GRANTED");
   //Serial.println("Access Granted");

  }
  else
  {
```

```
  lcd.setCursor(0,1);
  lcd.print("ACCESS DENIED");
  lcd.setCursor(0,3);
  lcd.print("NOT VERIFIED");
  // Serial.println("Access Denied");
  }
 entry_control=false;
 tag_status=false;
 }
}
```

22

Global System for Mobile-Based Emergency System

22.1 Introduction

This chapter explains the global system for mobile (GSM)-based emergency alert system in case of fire. The system has two sections: (1) sensor node and (2) mobile phone (to receive message in case of fire). The sensor node comprises Arduino Uno, power supply, fire sensor, LCD, and GSM modem. A fire sensor is connected to detect fire in the surrounding. If fire happens, the system will generate an alert message and that is sent to the registered mobile number through the GSM modem. Figure 22.1 shows the block diagram of the system.

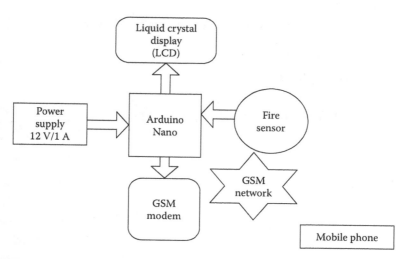

FIGURE 22.1
Block diagram of the system.

To design the system, following components are required (Table 22.1).

TABLE 22.1

Components List for a Sensor Node

Component/Specification	Quantity
Power supply/+12 V/1 A	1
Arduino Nano	1
Fire sensor	1
Fire sensor patch	1
GSM modem	1
MAX232 board (if GSM modem is with max232)	1
LCD (20 * 4)	1
LCD patch	1
2.4 GHz RF modem	1
2.4 GHz RF modem patch	1
Connecting wires (M–M, M–F, F–F)	20 each
Zero-size PCB or bread board or a designed PCB	1

22.2 Circuit Diagram

To develop the system, connect the components as follows:

Arduino and Flame sensor
- Arduino GND-Module GND
- Arduino +5 V-Module +
- Arduino digital pin 13-Module digital out pin

Arduino and GSM modem

- Arduino GND-Module GND
- Arduino +5 V-Module +
- Arduino Tx Pin-Rx pin of module
- Arduino Rx Pin-Tx pin of module

Arduino and LCD

- Arduino digital pin 12-RS pin(4) of LCD
- Arduino digital pin GND-RW pin(5) of LCD
- Arduino digital pin 11-E pin(6) of LCD
- Arduino digital pin 10-D4 pin(11) of LCD
- Arduino digital pin 9-D5 pin(12) of LCD
- Arduino digital pin 8-D6 pin(13) of LCD
- Arduino digital pin 7-D7 pin(14) of LCD

Figure 22.2 shows the circuit diagram for the system, connecting all the components as per connections.

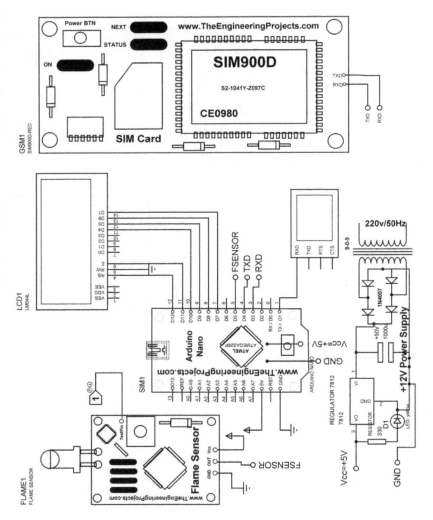

FIGURE 22.2

A circuit diagram for the system.

22.3 Program

22.3.1 Program to Write the Message

```
#include <LiquidCrystal.h>
LiquidCrystal lcd(12, 11, 10, 9, 8,7);

#include <SoftwareSerial.h>
SoftwareSerial mySerial(3, 4);
#define FIRE_SENSOR_button 5

String number ="9837043685"; // Add the 10-Digit Mobile Number
to which message / call is to be made,by replacing the 0's
void setup()
{
 lcd.begin(20, 4);
 Serial.begin(9600);
 mySerial.begin(9600);
 pinMode(FIRE_SENSOR_button,INPUT_PULLUP);

 lcd.setCursor(0, 0);
 lcd.print("GSM Based emergency ");
 lcd.setCursor(0, 1);
 lcd.print("system at UPES.... ");
 delay(2000);
}
void loop()
{
 int FIRE_STATUS_READ=digitalRead(FIRE_SENSOR_button);
 if (digitalRead(FIRE_STATUS_READ)==LOW) // Check if the sms
 key is being pressed
  {
   lcd.clear();
   mySerial.println("AT+CMGF=1"); // Set the Mode as Text Mode
   lcd.setCursor(0, 2);
   lcd.print("AT+CMGF=1");
   delay(150);
   mySerial.println("AT+CMGS=\"+919837043685\"\r"); // Specify
   the Destination number in international format by replacing
   the 0's
   lcd.clear();
   lcd.setCursor(0, 2);
   lcd.print("AT+CMGS=\"+919837043685\"\r");
   delay(150);
   lcd.clear();
   mySerial.print("message Send"); // Enter the message
   lcd.setCursor(0, 2);
   lcd.print("message Send");
```

```
  delay(150);
  lcd.clear();
  mySerial.write((byte)0x1A); // End of message character
  0x1A : Equivalent to Ctrl+z
  lcd.setCursor(0, 2);
  lcd.print("cntrl+Z");
  delay(50);
  mySerial.println();
  delay(50);
  lcd.setCursor(0, 3);
  lcd.print("FIRE DETECTED");
  delay(5000);

  }
 else
  {
  lcd.clear();
  lcd.setCursor(0, 3);
  lcd.print("NO FIRE");
  delay(20);
  }

}
```

22.3.2 Program to Read the Message

```
#include <LiquidCrystal.h>
LiquidCrystal lcd(12, 11, 5, 4, 3, 2);
char Rec_Data=0;
byte msg_flg=0;
byte msg_flag_2=0;
byte call_flag=0;
byte reply_flag=0;
byte delete_flag=0;
byte d_count=0;
byte i=0;
byte j=0;
char number[15];
char message[160];
void modem_initialization(void);
void gsm_send_message(void);
void clear_rx_buffer(void);
void setup()
{
 delay(1000);
 Serial.begin(9600);//start serial port at baud rate of 9600
 lcd.begin(16,2);
 lcd.clear();
 lcd.setCursor(0,0);
```

```
lcd.print("Arduino with GSM");
lcd.setCursor(0,1);
lcd.print(" UPES ");
delay(1000);
delay(1000);
lcd.clear();
lcd.setCursor(0,0);
lcd.print("GSM Initialising     ");
modem_initialization();
lcd.clear();
lcd.setCursor(0,0);
lcd.print("GSM Initialised      ");
}
void loop()
{
gsm_read_message();//wait for message or call
if( reply_flag == 1 )//only if valid message received
  {
  lcd.clear();
  lcd.setCursor(0,0);lcd.print(number);
  lcd.setCursor(0,1);lcd.print(message);
  gsm_send_message();
  }
}
void gsm_send_message(void)
{
byte send_flag=1;
byte msg_flg=0;
char rec_data=0;
  delay(300);
  clear_rx_buffer();//clearing receiving buffer
  Serial.print("AT+CMGS=");//Command to send message ,
  (AT+CMGS="+91phone number"<enter>)
  Serial.print('"');
  Serial.print(number);
  Serial.println('"');
  while(send_flag == 1)
  {
  while(Serial.available() > 0)
   {
    rec_data = Serial.read();
      if(rec_data == '>')// '>'(greater than symbol) is recieved
      while message sending,after AT+CMGD="number" command
        {
          delay(400);//Delay for GSM module become ready
          Serial.print(" MESSAGE RECEIVED AND DISPLAYED IN
          LCD,THANK YOU,UPES");//Transmitting whole string to
          be sent as SMS
          Serial.write(26);//'ctrl+z' command
```

```
        delay(300);//Delay for GSM module become ready
        msg_flg=0;
    }
else if( rec_data == '+' && msg_flg == 0 )// '+' is
recieved
    msg_flg = 1 ;
else if( rec_data == 'C' && msg_flg == 1 )// 'C' is
recieved
    msg_flg = 2 ;
else if( rec_data == 'M' && msg_flg == 2 )// 'M' is
recieved
    msg_flg = 3 ;
else if( rec_data == 'G' && msg_flg == 3 )// 'M' is
recieved
    msg_flg = 4 ;
else if ( rec_data == 'S' && msg_flg == 4 )//// 'S' is
recieved after a sucessful message ("+CMSGS: ")
    msg_flg = 5 ;
else if ( rec_data == ':' && msg_flg == 5 )// ':' is
recieved
    msg_flg = 6 ;
else if ( rec_data == 0x0D && msg_flg == 6 )//Carriage
return
    msg_flg = 7 ;
else if ( rec_data == 0x0A && msg_flg == 7 )//Line feed
    msg_flg = 8 ;
else if ( rec_data == '0' && msg_flg == 8 )// '0'
(zero) is recieved
        {
            clear_rx_buffer();//clearing receiving buffer
            msg_flg=0;
            msg_flag_2=0;
            send_flag=0;
            reply_flag=0;
        }
else if ( rec_data == '4' && msg_flg == 0 )// '4'
(zero) is recieved
        {
        clear_rx_buffer();//clearing receiving buffer
        msg_flg=0;
        msg_flag_2=0;
        send_flag=0;
        reply_flag=0;
        }
else if( Rec_Data == '2' && msg_flg == 0 )// '2' is
recieved,incoming call
        {
        call_flag=1;
        delay(300);
```

```
                Serial.println("ATH");//Command for hang up
                incomming call,all calls must be filtered no
                need to answer a call
                }
        else if( Rec_Data=='0' && msg_flg == 0 &&
        call_flag == 1 )// '0' (zero) is recieved,call
        sucessfully rejected
                {
                msg_flg=0;
                call_flag=0;
                msg_flag_2=0;
                d_count=0;
                send_flag=0;
                reply_flag=0;
                }


        }
    }
}
void modem_initialization(void)
{
 char rec_data;
 byte network_status = 0 ;//network_status initialized as
 zero
 byte status_check = 0 ;//status_check initialized as zero
 byte gsm_cnt = 0;//gsm_cnt initialized as zero
 byte ok_flag=0;//ok_flag initialized as zero
 byte count=0;//count initialized as zero
 clear_rx_buffer();//clearing receiving buffer
 while(gsm_cnt < 5)// repeat entire loop until gsm_cnt less
 than seven
 {
     switch(gsm_cnt)
        {
        case 0: clear_rx_buffer();//clearing receiving buffer
            Serial.println("AT");// Attention command to wake
            up GSM modem
            delay(1000);
            break;
        case 1: clear_rx_buffer();//clearing receiving buffer
            Serial.println("ATE0");//Command for disable echo
            delay(1000);
            break;
        case 2: clear_rx_buffer();//clearing receiving buffer
            Serial.println("ATV0");// Command for numeric
            response after this '0'(zero) will be recieved
            instead of "OK"
            delay(1000);
            break;
        case 3: clear_rx_buffer();//clearing receiving buffer
```

```
            Serial.println("AT&W");// Command TO SAVE
            SETTINGS
            delay(1000);
            break;
        case 4: gsm_cnt = 5;// exit from the loop
            break;
        default :break;
      }
   while(Serial.available() > 0)
      {
        rec_data = Serial.read();
        if(rec_data == 'O' )// 'o' is recieved
           ok_flag = 1;
        else if(ok_flag == 1 && rec_data == 'K' )// 'K' is
        recieved
           {
             gsm_cnt ++;
             ok_flag = 0;
           }
        else if (rec_data == '0' )// '0'(zero) is recieved
        (numeric response)
           {
             gsm_cnt ++;
             ok_flag = 0;
           }
        else if (rec_data == '+' )// '+' is recieved message
           {
           clear_rx_buffer();//clearing receiving buffer

                if( gsm_cnt > 0 )
                        gsm_cnt --;
           }
      }
}
lcd.setCursor(0,0);
lcd.print("GSM Modem Found ");
gsm_cnt = 0;
ok_flag = 0;
while(network_status == 0 )//wait for PIN READY
      {
           if(status_check == 0 )
             {
               delay(1000);
               status_check = 1 ;
               clear_rx_buffer();//clearing receiving buffer
```

```
        Serial.println("AT+CPIN?"); //checkin PIN return
        ready(+CPIN: READY) with a valid simcard
        otherwise error
    }
    while(Serial.available() > 0)
    {
        rec_data = Serial.read();
        if( rec_data == '+' && status_check == 1 )
        // '+' is recieved
        status_check = 2 ;
        else if( rec_data == 'C' && status_check == 2 )
        // 'C' is recieved
        status_check = 3 ;
        else if( rec_data == 'P' && status_check == 3 )
        // 'P' is recieved
        status_check = 4 ;
        else if( rec_data == 'I' && status_check == 4 )
        // 'I' is recieved
        status_check = 5 ;
        else if( rec_data == 'N' && status_check == 5 )
        // 'N' is recieved
        status_check = 6 ;
        else if( rec_data == ':' && status_check == 6 )
        // ':' is recieved
        status_check = 7 ;
        else if( rec_data == ' ' && status_check == 7 )
        // ' ' is recieved
        status_check = 8 ;
        else if( rec_data == 'R' && status_check == 8 )
        // 'R' is recieved
        status_check = 9 ;
        else if( rec_data == 'E' && status_check == 9 )
        // 'E' is recieved
        status_check = 10 ;
        else if( rec_data == 'A' && status_check == 10 )
        // 'A' is recieved
        status_check = 11 ;
        else if( rec_data == 'D' && status_check == 11 )
        // 'D' is recieved
        status_check = 12 ;
        else if( rec_data == 'Y' && status_check == 12 )
        // 'Y' is recieved
        status_check = 13 ;
        else if( rec_data == 0X0D && status_check == 13 )
        //Carriage return
```

```
                  status_check = 14 ;
              else if( rec_data == 0X0A && status_check == 14 )
              //Line Feed
                 status_check = 15 ;
              else if( rec_data == '0' && status_check == 15 )
              // '0' is recieved
                  {
                      clear_rx_buffer();//clearing receiving
                      buffer
                      status_check = 0 ;
                      network_status=1;//goto next step
                  }
              else if( rec_data != 'R' && status_check == 8 )
              //+CPIN: NOT READY
                  {
                      clear_rx_buffer();//clearing receiving buffer
                      status_check = 0 ;
                      network_status=0;//repeat current step
                  }
              else if( rec_data == 'M' && status_check == 3 )
              // in case of any message
                  {
                      clear_rx_buffer();//clearing receiving
                      buffer
                      status_check = 0 ;
                      network_status=0;//repeat current step
                  }

              }

          }
  lcd.setCursor(0,0);
  lcd.print("CPIN: Ready ");
  while(network_status == 1 )//wait for SIM network
  registration
          {
              if(status_check == 0 )
                {
                   delay(1000);
                   status_check = 1 ;
                   clear_rx_buffer();//clearing receiving buffer
                   Serial.println("AT+CREG?"); //checking for SIM
                   card registration,if registerd "+CREG: 0,1"
                   will receive
                }
              while(Serial.available() > 0)
                {
                   rec_data = Serial.read();
                      if( rec_data == '+' && status_check == 1 )
                      // '+' is recieved
```

```
        status_check = 2 ;
    else if( rec_data == 'C' && status_check == 2 )
    // 'C' is recieved
        status_check = 3 ;
    else if( rec_data == 'R' && status_check == 3 )
    // 'R' is recieved
        status_check = 4 ;
    else if( rec_data == 'E' && status_check == 4 )
    // 'E' is recieved
        status_check = 5 ;
    else if( rec_data == 'G' && status_check == 5 )
    // 'G' is recieved
        status_check = 6 ;
    else if( rec_data == ':' && status_check == 6 )
    // ':' is recieved
        status_check = 7 ;
    else if( rec_data == ' ' && status_check == 7 )
    // ' ' is recieved
        status_check = 8 ;
    else if( rec_data == '0' && status_check == 8 )
    // '0' is recieved
        status_check = 9 ;
    else if( rec_data == ',' && status_check == 9 )
    // ',' is recieved
        status_check = 10 ;
    else if( rec_data == '1' && status_check == 10 )
    // '1' is recieved
        status_check = 11 ;
    else if( rec_data == 0X0D && status_check == 11 )
    //Carriage return
        status_check = 12 ;
    else if( rec_data == 0X0A && status_check == 12 )
    // Line Feed
        status_check = 13 ;
    else if( rec_data == '0' && status_check == 13 )
    // '0' is recieved
            {
                clear_rx_buffer();//
                  clearing receiving buffer
                status_check = 0 ;
                network_status=2;//goto next step
            }
    else if( rec_data != '1' && status_check == 10 )
    // +CREG: 0,2 not registered
```

```
                    {
                      clear_rx_buffer();//
                        clearing receiving buffer
                      status_check = 0 ;
                      network_status=1;//repeat current step
                    }
             else if( rec_data == 'M' && status_check == 3 )
             // in case of any message
                    {
                      clear_rx_buffer();//clearing receiving
                        buffer
                      status_check = 0 ;
                      network_status=1;//repeat current step
                    }

          }

      }
 gsm_cnt=0;
lcd.setCursor(0,0);
lcd.print("Registration Ok        ");
while(gsm_cnt < 4)// repeat entire loop until gsm_cnt less
than nine
 {
    switch(gsm_cnt)
      {
        case 0: clear_rx_buffer();//clearing receiving buffer
                Serial.println("AT+CMGF=1");// Attention
                command to wake up GSM modem
                delay(1000);
                break;
        case 1: clear_rx_buffer();//clearing receiving buffer
                Serial.println("AT+CNMI=2,1,0,0,0");
                //Command to configure new message indication
                delay(1000);
                break;
        case 2: clear_rx_buffer();//clearing receiving buffer
                Serial.println("AT+CMGD=1,4");// Command
                to delete all received messages
                delay(1000);
                break;
        case 3: gsm_cnt = 4;// exit from the loop
                break;
        default :break;
      }
    while(Serial.available() > 0)
      {
        rec_data = Serial.read();
```

```
        if (rec_data == '0' )// '0'(zero) is recieved (numeric
        response)
            {
                gsm_cnt ++;
            }
        else if (rec_data == '+'  )// '+' recieved ,before
        "AT+CIICR" command,may be any message
            {
            clear_rx_buffer();//clearing receiving buffer
                if( gsm_cnt > 0 )
                        gsm_cnt --;
            }

        else if (rec_data == '4' && gsm_cnt > 2)// '4' recieved
        (error),in gprs initialisation commands
            {
            clear_rx_buffer();//clearing receiving buffer
            gsm_cnt = 4;
            }
        }
    }
    gsm_cnt = 0;
}
void clear_rx_buffer(void)
{
char rec_data=0;
while(Serial.available() > 0)
    rec_data = Serial.read();
}
void gsm_read_message(void)
{
while(Serial.available() > 0)
    {
        Rec_Data = Serial.read();
        if( Rec_Data == '+' && msg_flg == 0 )
        // '+' is recieved
          msg_flg=1;
        else if( Rec_Data == 'C' && msg_flg == 1 )
        // 'C' is recieved
          msg_flg=2;
        else if( Rec_Data == 'M' && msg_flg == 2 )
        // 'M' is recieved
          msg_flg=3;
        else if( Rec_Data == 'T' && msg_flg == 2 )
        // 'M' is recieved
          msg_flg=3;
```

```
else if( Rec_Data == 'I' && msg_flg == 3 )
// ':' is recieved (INCOMING MESSAGE )
    {
      clear_rx_buffer();//clearing receiving buffer
      Rec_Data =0;
      msg_flg=0;
      delay(300);//Delay for GSM module become ready
      array_clear();//Clear both message and number
      arrays
      Serial.println("AT+CMGR=1");//Command for message
      read from location one
      i=0;j=0;
    }
else if( Rec_Data== 'G' && msg_flg == 3 )
// 'M' is recieved
  msg_flg=4;
else if( Rec_Data== 'R' && msg_flg == 4 )
// 'r' is recieved,Readind message
  msg_flg=5;
else if( Rec_Data== ':' && msg_flg == 5 )
// ':' is recieved
  msg_flg=6;
else if( Rec_Data == '"' && msg_flg == 6  )
// Counting double quotes
  d_count++;
else if( Rec_Data != '"' && d_count==3 && msg_
flg == 6) // Saving number to message
array from between third and fourth double quotes
  number[i++] = Rec_Data;
else if( Rec_Data== 0X0D && msg_flg==6)//
Carriage return
  msg_flg=7;
else if( Rec_Data== 0X0A && msg_flg==7)//Line feed
  msg_flg=8;
else if( Rec_Data=='*' && d_count>=5 &&  msg_
flg == 8 ) //Start symbol '*' is recieved
  msg_flag_2=1;
else if( Rec_Data!='#' && msg_flag_2 == 1 && msg_
flg == 8 && Rec_Data!=0X0D && Rec_Data!=0X0A )
//Data between start and stop symbols were saved
to message array
  message[j++]=Rec_Data;
else if( Rec_Data=='#' && msg_flag_2 == 1 &&  msg_
flg == 8)//Stop symbol '#' is recieved
    {
      msg_flag_2 = 2;
      number[i]='\0';
      message[j]='\0';
      i=0;j=0;
    }
```

```
else if(Rec_Data == 0X0D && msg_flg == 8)
//Carriage return (end of message)
    {
    clear_rx_buffer();//clearing receiving buffer
    msg_flg=0;
    delete_flag=1;
    delay(300);//Delay for GSM module become ready
    Serial.println("AT+CMGD=1,4"); // Command for
    delete all messages in SIM card
    }
else if( Rec_Data == '0' && msg_flg == 0 && delete_
flag == 1 )//Response for delete command ,sucess all
messages were deleted
    {
     if( msg_flag_2 == 2 )
       reply_flag = 1;//If the message have start
       and stop symbols it must be checked otherwise
       no need to check the recieved message
     else
       reply_flag = 0;//No need to check
       recieved message
     msg_flg=0;
     msg_flag_2=0;
     d_count=0;
     delete_flag=0;
    }
else  if( Rec_Data == '2' && msg_flg == 0 )// '2' is
recieved,incoming  call
    {
    call_flag=1;
    delay(300);
    Serial.println("ATH");//Command  for  hang  up
    incomming  call,all  calls  must  be  filtered
    no  need  to  answer  a  call
    }
else  if(  Rec_Data=='0'  &&  msg_flg  ==  0  &&
call_flag  ==  1  )//  '0'  (zero)  is  recieved,call
sucessfully  rejected
    {
    msg_flg=0;
    call_flag=0;
    msg_flag_2=0;
    d_count=0;
    reply_flag=0;
    }
else  if(  (Rec_Data  ==  '3'  ||  Rec_Data  ==  '4'  ||
Rec_Data  ==  '7'  )  &&  msg_flg  ==  0  )//incase
of  any  error,nocarrier  and  busy
    {
```

```
                    msg_flg=0;
                    call_flag=0;
                    msg_flag_2=0;
                    d_count=0;
                    reply_flag=0;
                    }
         }
}

void  array_clear(void)
{
byte  k=0;
    for(k=0;k<=15;k++)
    number[k]='\0';
    for(k=0;k<=63;k++)
    message[k]='\0';
}
```

22.4 Proteus Simulation Model

Figures 22.3 and 22.4 shows the Proteus simulation model for interfacing the GSM modem with the Arduino. Load the program in the Arduino and check the working of the circuit. The LCD displays the status of the system.

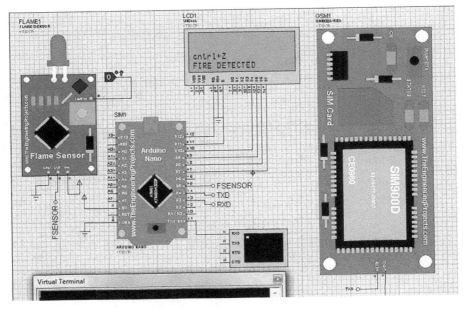

FIGURE 22.3
Proteus simulation model of the system showing "FIRE DETECTED."

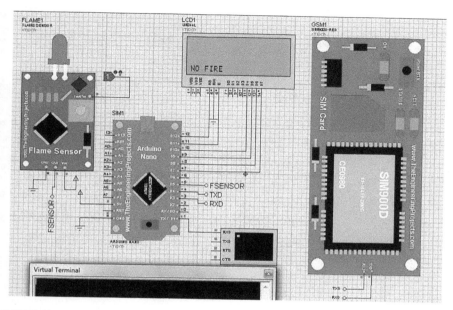

FIGURE 22.4
Proteus simulation model of the system showing "NO FIRE."

23

Coordinate Display System Using GPS

23.1 Introduction

This chapter explains the global positioning system (GPS)-based coordinate display system. The system has two sections: (1) transmitter section and (2) receiver section. The transmitter section comprises the Arduino Uno, power supply, LCD, GPS module, and the RF modem. GPS is connected to detect the coordinates of node location. The receiver section comprises the Arduino Uno, power supply, LCD, RF modem, and the terminal v1.9b. The RF modem is used for short-distance wireless communication and the terminal v1.9b is used to display the coordinates with the PC. It can be used as a localization system in the robots. Figure 23.1 shows the snapshot of the GPS module. Figure 23.2 shows the block diagram of the system.

To design the system, following components are required (Tables 23.1 and 23.2).

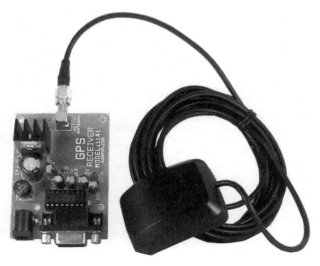

FIGURE 23.1
A GPS module.

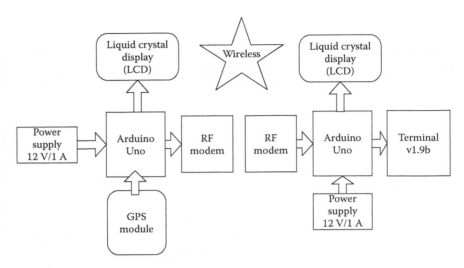

FIGURE 23.2
Block diagram of the system.

TABLE 23.1

Components List for a Transmitter Section

Component/Specification	Quantity
Power supply/+12 V/1 A	1
Arduino Uno	1
GPS module	1
LCD (20 * 4)	1
LCD patch	1
2.4 GHz RF modem	1
2.4 GHz RF modem patch	1
Connecting wires (M–M, M–F, F–F)	20 each
Zero-size PCB or bread board or a designed PCB	1

TABLE 23.2

Components List for a Receiver Section

Component/Specification	Quantity
Power supply/+12 V/1 A	1
Arduino Uno	1
LCD (20 * 4)	1
LCD patch	1
2.4 GHz RF modem	1
2.4 GHz RF modem patch	1
PC (terminal v1.9b)	1
Connecting wires (M–M, M–F, F–F)	20 each
Zero-size PCB or bread board or a designed PCB	1

23.2 Circuit Diagram

To develop the system, connect the components as follows:

23.2.1 Transmitter Section

Arduino and GPS

- Arduino GND-Module GND
- Arduino +5 V-Module +
- Arduino Rx Pin-data out pin of GPS

Arduino and RF modem

- Arduino pin0(TX)-RX pin of RF modem
- Arduino pin0(RX)-Tx pin of RF modem
- Arduino+5 V-+Vcc of RF modem
- Arduino Ground-GND of RF modem

Arduino and LCD

- Arduino digital pin 12-RS pin(4) of LCD
- Arduino digital pin GND-RW pin(5) of LCD
- Arduino digital pin 11-E pin(6) of LCD
- Arduino digital pin 10-D4 pin(11) of LCD
- Arduino digital pin 9-D5 pin(12) of LCD
- Arduino digital pin 8-D6 pin(13) of LCD
- Arduino digital pin 7-D7 pin(14) of LCD

23.2.2 Receiver Section

Arduino and RF modem

- Arduino pin0(TX)-RX pin of RF modem
- Arduino pin0(RX)-Tx pin of RF modem
- Arduino+5 V-+Vcc of RF modem
- Arduino Ground-GND of RF modem

Arduino and LCD

- Arduino digital pin 12-RS pin(4) of LCD
- Arduino digital pin GND-RW pin(5) of LCD
- Arduino digital pin 11-E pin(6) of LCD
- Arduino digital pin 10-D4 pin(11) of LCD
- Arduino digital pin 9-D5 pin(12) of LCD
- Arduino digital pin 8-D6 pin(13) of LCD
- Arduino digital pin 7-D7 pin(14) of LCD

Figure 23.3 shows the circuit diagram for the system, connecting all the components as per connections.

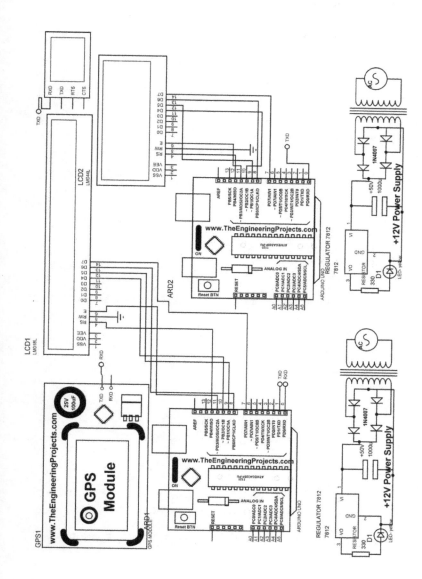

FIGURE 23.3
A circuit diagram of the system.

23.3 Program

23.3.1 Transmitter Section

```
#include <TinyGPS.h>
#include <LiquidCrystal.h>
LiquidCrystal lcd(12, 11, 10, 9, 8, 7);
TinyGPS gps;   //Creates a TinyGPS object

void setup()
{
  Serial.begin(9600);
  lcd.begin(40,2);
  lcd.print("GPS DISPLAY SYSTEM");
  delay(1000);
  lcd.clear();
}

void loop()
{
  bool newData = false;
  unsigned long chars;
  unsigned short sentences, failed;

  for (unsigned long start = millis(); millis() - start < 1000;)
  {
    while (Serial.available())
    {
    char c = Serial.read();
    if (gps.encode(c))
        newData = true;
    }
  }

  if (newData)        //If newData is true
  {
    float flat, flon;
    unsigned long age;
    gps.f_get_position(&flat, &flon, &age);
    Serial.print(flat == TinyGPS::GPS_INVALID_F_ANGLE
    ? 0.0 : flat, 6);
    Serial.println(flon == TinyGPS::GPS_INVALID_F_ANGLE
    ? 0.0 : flon, 6);

    lcd.setCursor(0,0);
    lcd.print("Latitude = ");
    lcd.print(flat == TinyGPS::GPS_INVALID_F_ANGLE
    ? 0.0 : flat, 6);
```

```
      lcd.setCursor(0,1);
      lcd.print(" Longitude = ");
      lcd.print(flon == TinyGPS::GPS_INVALID_F_ANGLE
      ? 0.0 : flon, 6);
    }

  Serial.println(failed);
}
```

23.3.2 Receiver Section

```
#include <LiquidCrystal.h>
LiquidCrystal lcd(12, 11, 10, 9, 8, 7);
String inputString = "";          // choose blank string to
hold data
boolean stringComplete = false;
void setup()
{
Serial.begin(9600);
lcd.begin(20,4);
inputString.reserve(200);
lcd.setCursor(0,0);
lcd.print("GPS Cordinate");
lcd.setCursor(0,1);
lcd.print("Capturing SYSTEM");
delay(1000);
lcd.clear();
}

void loop()
{

  if (stringComplete)

  {
    lcd.clear();
    Serial.println(inputString);
    lcd.setCursor(0,2);
    lcd.print("LATITUDE:");
    lcd.print(inputString[0]);
    lcd.print(inputString[1]);
    lcd.print(inputString[2]);
    lcd.print(inputString[3]);
    lcd.print(inputString[4]);
    lcd.print(inputString[5]);
    lcd.print(inputString[6]);
    lcd.print(inputString[7]);
    lcd.print(inputString[8]);
```

```
    lcd.setCursor(0,3);
    lcd.print("LONGITUDE:");
    lcd.print(inputString[9]);
    lcd.print(inputString[10]);
    lcd.print(inputString[11]);
    lcd.print(inputString[12]);
    lcd.print(inputString[13]);
    lcd.print(inputString[14]);
    lcd.print(inputString[15]);
    lcd.print(inputString[16]);
    lcd.print(inputString[17]);
    lcd.print(inputString[18]);
    lcd.print(inputString[19]);

        // clear the string:

    if(inputString[0]==0x0A)
    {
    lcd.setCursor(0,2);
    lcd.print("LATITUDE:");
    lcd.print(inputString[1]);
    lcd.print(inputString[2]);
    lcd.print(inputString[3]);
    lcd.print(inputString[4]);
    lcd.print(inputString[5]);
    lcd.print(inputString[6]);
    lcd.print(inputString[7]);
    lcd.print(inputString[8]);
    lcd.print(inputString[9]);
    lcd.setCursor(0,3);
    lcd.print("LONGITUDE:");
    lcd.print(inputString[10]);
    lcd.print(inputString[11]);
    lcd.print(inputString[12]);
    lcd.print(inputString[13]);
    lcd.print(inputString[14]);
    lcd.print(inputString[15]);
    lcd.print(inputString[16]);
    lcd.print(inputString[17]);
    lcd.print(inputString[18]);
    lcd.print(inputString[19]);
    lcd.print(inputString[20]);
    }
    inputString = "";
    stringComplete = false;
  }
  delay(200);
}
```

```
void serialEvent()
{
  while (Serial.available())
  {
    char inChar = (char)Serial.read(); // read serial
    inputString += inChar;// store byte in string
    if (inChar == 0x0D)
    {
      stringComplete = true;
    }
  }
}
```

23.4 Proteus Simulation Model

Figure 23.4 shows the Proteus simulation model for interfacing the GPS module with the Arduino. Load the program in the Arduino and check the working of the circuit. LCD displays the status of the system. The virtual terminal also displays the coordinate values.

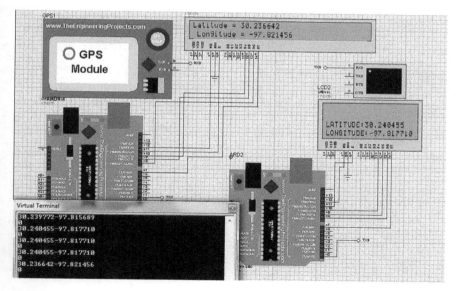

FIGURE 23.4
Proteus simulation model for the system.

24

Fingerprint-Based Attendance System

24.1 Introduction

This chapter discusses about the fingerprint-based attendance system. The fingerprint sensor can read different fingerprints and store them in its flash memory. It can perform functions such as ADD, Empty, or Search and return to the stored ID. These functions can be performed with three onboard switches. Fingerprint of each student is first identified and stored. The fingerprint of student is matched with the pre stored data, to authenticate the presence of a student. The system comprises of Arduino Nano, power supply, and fingerprint sensor.

24.1.1 Types of Function

There are three functions for the fingerprint sensor:

Add (Enroll) function: It adds a fingerprint to the database, and return a byte of the added ID. Return values are from 0x00 to 0xFE. In case of error or if no finger is placed, return code is 0xFF.

Search function: If a finger is put and the search function is called, it returns a matching ID if found in its existing memory. The return values are from 0x00 to 0xFE. In case of error or if no finger is placed, return code is 0xFF.

Empty function: Empty all the fingerprint data that are stored. After executing this function, it is 0xCC as OK or 0xFF as error.

Figure 24.1 shows the block diagram of the system.
To design the system, following components are required (Table 24.1).

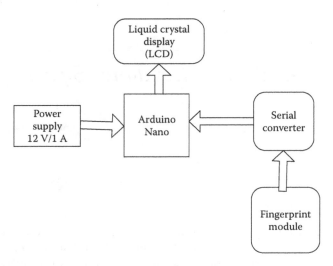

FIGURE 24.1
Block diagram of the system.

TABLE 24.1

Components List for a System

Component/Specification	Quantity
Power supply/+12 V/1 A	1
Arduino Nano	1
Fingerprint module	1
LCD (20 * 4)	1
LCD patch	1
Connecting wires (M–M, M–F, F–F)	20 each
Zero-size PCB or bread board or a designed PCB	1

24.2 Circuit Diagram

To develop the system, connect the components as follows:

Arduino and fingerprint sensor

- Arduino pin2-Search pin of fingerprint module
- Arduino pin 0(RX)-Tx out pin of module
- Arduino +5 V pin-+Vcc of module
- Arduino GND pin-GND of module

Arduino and LCD

- Arduino digital pin 12-RS pin(4) of LCD
- Arduino digital pin GND-RW pin(5) of LCD
- Arduino digital pin 11-E pin(6) of LCD
- Arduino digital pin 10-D4 pin(11) of LCD
- Arduino digital pin 9-D5 pin(12) of LCD
- Arduino digital pin 8-D6 pin(13) of LCD
- Arduino digital pin 7-D7 pin(14) of LCD

Figure 24.2 shows the circuit diagram of the system.

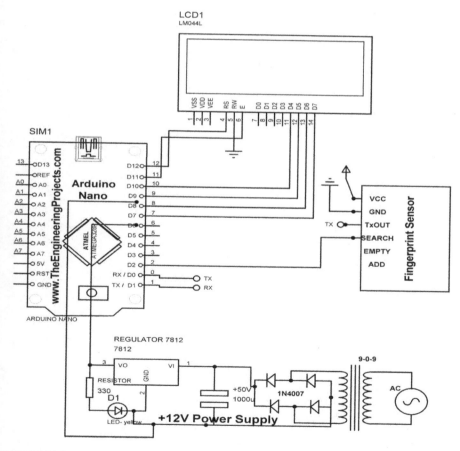

FIGURE 24.2
A circuit diagram of the system.

24.3 Program

24.3.1 Finger Print Circuit 'Test Program'

```
#include <LiquidCrystal.h>
LiquidCrystal lcd(13, 12, 11, 10, 9, 8);

int FF_PIN1=7;
int x_PIN1=6;
int y_PIN1=5;
int z_PIN1=4;
int w_PIN1=3;
int p_PIN1=2;
void setup()
{
    lcd.begin(20,4);
    Serial.begin(9600);
     pinMode(FF_PIN1,INPUT_PULLUP);
     pinMode(x_PIN1,INPUT_PULLUP);
     pinMode(y_PIN1,INPUT_PULLUP);
     pinMode(z_PIN1,INPUT_PULLUP);
     pinMode(w_PIN1,INPUT_PULLUP);
     pinMode(p_PIN1,INPUT_PULLUP);
    lcd.setCursor(0, 0);
    lcd.print("Fingerprint based");
    lcd.setCursor(0, 1);
    lcd.print("Security System");

}

void loop()
{
  int FF_PIN=digitalRead(FF_PIN1);
   int x_PIN=digitalRead(x_PIN1);
    int y_PIN=digitalRead(y_PIN1);
     int z_PIN=digitalRead(z_PIN1);
      int w_PIN=digitalRead(w_PIN1);
       int p_PIN=digitalRead(p_PIN1);
  if(FF_PIN==LOW)
  {

  lcd.clear();
  lcd.setCursor(0, 2);
  lcd.print("send FF");
  Serial.write(0xFF);
  delay(20);
  }
```

```
if(x_PIN==LOW)
{

lcd.clear();
lcd.setCursor(0, 2);
lcd.print("send 00");
Serial.write(0x00);
delay(20);
}
if(y_PIN==LOW)
{

lcd.clear();
lcd.setCursor(0, 2);
lcd.print("send 01");
Serial.write(0x01);
delay(20);
}
if(z_PIN==LOW)
{

lcd.clear();
lcd.setCursor(0, 2);
lcd.print("send 02");
Serial.write(0x02);
delay(20);
}
if(w_PIN==LOW)
{

lcd.clear();
lcd.setCursor(0, 2);
lcd.print("send 03");
Serial.write(0x03);
delay(20);
}
if(p_PIN==LOW)
{

lcd.clear();
lcd.setCursor(0, 2);
lcd.print("send 04");
Serial.write(0x04);
delay(20);
}
}
```

24.3.2 Main Program

```
#include <LiquidCrystal.h>
LiquidCrystal lcd(13, 12, 11, 10, 9, 8);

int SEARCH_PIN=7;

void setup()
{
  lcd.begin(20,4);
  Serial.begin(9600);
  pinMode(SEARCH_PIN,OUTPUT);
  digitalWrite(SEARCH_PIN,HIGH);
  lcd.setCursor(0, 0);
  lcd.print("Fingerprint based");
  lcd.setCursor(0, 1);
  lcd.print("Security System");
}

void loop()
{
digitalWrite(SEARCH_PIN,HIGH);
delay(500);
digitalWrite(SEARCH_PIN,LOW);
delay(500);
  int DATA_FINGERPRINT=Serial.read();

 if(DATA_FINGERPRINT==0xFF)
 {
  lcd.clear();
  lcd.setCursor(0, 0);
  lcd.print("show your ");
  lcd.setCursor(0, 1);
  lcd.print("right thomb");
  delay(20);
  }
 if(DATA_FINGERPRINT==0x00)
 {
  lcd.clear();
  lcd.setCursor(0, 0);
  lcd.print("Authenticate ");
  lcd.setCursor(0, 1);
  lcd.print("THANKS RAJESH");
  delay(20);
 }
 if(DATA_FINGERPRINT==0x01)
 {
  lcd.clear();
  lcd.setCursor(0, 0);
```

```
lcd.print("Authenticate ");
lcd.setCursor(0, 1);
lcd.print("THANKS BHUPENDRA");
delay(20);
}
if(DATA_FINGERPRINT==0x02)
{
 lcd.clear();
 lcd.setCursor(0, 0);
 lcd.print("Authenticate ");
 lcd.setCursor(0, 1);
 lcd.print("THANKS ANITA");
 delay(20);
}
if(DATA_FINGERPRINT==0x03)
{
 lcd.clear();
 lcd.setCursor(0, 0);
 lcd.print("Authenticate ");
 lcd.setCursor(0, 1);
 lcd.print("THANKS CHOUDHURY");
 delay(20);
}
if(DATA_FINGERPRINT==0x04)
{
 lcd.clear();
 lcd.setCursor(0, 0);
 lcd.print("Authenticate ");
 lcd.setCursor(0, 1);
 lcd.print("THANKS XXXXXXX");
 delay(20);
}


}
```

25

Wireless Irrigation System for Agricultural Field

25.1 Introduction

This chapter elaborates the wireless irrigation system for the agricultural field. The idea is to measure the soil moisture of field along with the surrounding temperature. On the basis of the data received from the sensors, the motor is made "ON" or "OFF" to irrigate the fields, as per the requirement of the crop. The system is designed in two sections: (1) Sensor node and (2) Remote control. Sensor node comprises of the Arduino Uno, power supply, soil moisture sensor, temperature sensor, LCD, load (DC motor), and the RF modem. The remote control comprises of Arduino Uno, power supply, LCD, switch array (to send commands to operate the load), and RF modem. All the parameters can also be displayed on LabVIEW GUI. Figure 25.1 shows the block diagram of the system.

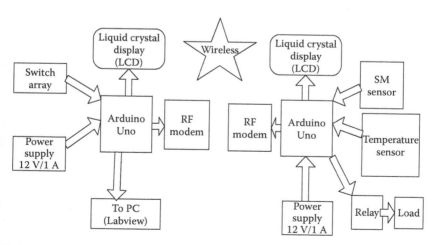

FIGURE 25.1
Block diagram of the system.

To design the system, following components are required (Tables 25.1 and 25.2).

TABLE 25.1

Components List for a Remote Control

Component/Specification	Quantity
Power supply/+12 V/1 A	1
Arduino Uno	1
Switches	4
LCD (20 * 4)	1
LCD patch	1
2.4 GHz RF modem	1
2.4 GHz RF modem patch	1
Connecting wires (M–M, M–F, F–F)	20 each
Zero-size PCB or bread board or a designed PCB	1

TABLE 25.2

Components List for a Sensor Node

Component/Specification	Quantity
Power supply/+12 V/1 A	1
Arduino Uno	1
Soil moisture sensor	1
Temperature sensor	1
DC motor	1
Relay	1
2N2222	1
L293D	1
LCD (20 * 4)	1
LCD patch	1
2.4 GHz RF modem	1
2.4 GHz RF modem patch	1
Connecting wires (M–M, M–F, F–F)	20 each
Zero-size PCB or bread board or a designed PCB	1

25.2 Circuit Diagram

To develop the system, connect the components as follows:

25.2.1 Remote Control

Arduino and Push Button

- Arduino pin7-one terminal of 'Mode selection' push button
- Arduino pin6-one terminal of 'Motor ON' button
- Arduino pin5-one terminal of 'Motor OFF' button
- Other terminal of all push buttons-GND

Arduino and RF modem

- Arduino pin0(TX)-RX pin of RF modem
- Arduino pin0(RX)-Tx pin of RF modem
- Arduino+5 V-+Vcc of RF modem
- Arduino Ground-GND of RF modem

Arduino and LCD

- Arduino digital pin 13-RS pin(4) of LCD
- Arduino digital pin GND-RW pin(5) of LCD
- Arduino digital pin 12-E pin(6) of LCD
- Arduino digital pin 11-D4 pin(11) of LCD
- Arduino digital pin 10-D5 pin(12) of LCD
- Arduino digital pin 9-D6 pin(13) of LCD
- Arduino digital pin 8-D7 pin(14) of LCD

25.2.2 Sensor Node

LM35 sensor connection

- Arduino GND-Module GND
- Arduino +5 V-Module +
- Arduino A0 pin-data out pin of sensor

Soil Moisture sensor connection
- Arduino GND-Module GND
- Arduino +5 V-Module +
- Arduino A1 pin-data out pin of sensor

2N2222, Relay, and Arduino connection
- Arduino pin 7-base of 2N2222
- Collector of 2N2222-L2 end relay
- L1 end of relay-+12 V of battery or power supply
- COM pin of relay-one end of AC motor
- NO pin of relay-other end of AC motor

Arduino and RF modem
- Arduino pin1(TX)-RX pin of RF modem
- Arduino pin0(RX)-Tx pin of RF modem
- Arduino+5 V-+Vcc of RF modem
- Arduino Ground-GND of RF modem

Arduino and LCD
- Arduino digital pin 12-RS pin(4) of LCD
- Arduino digital pin GND-RW pin(5) of LCD
- Arduino digital pin 11-E pin(6) of LCD
- Arduino digital pin 5-D4 pin(11) of LCD
- Arduino digital pin 4-D5 pin(12) of LCD
- Arduino digital pin 3-D6 pin(13) of LCD
- Arduino digital pin 2-D7 pin(14) of LCD

Figure 25.2 shows the circuit diagram for the system, connecting all the components as per connections.

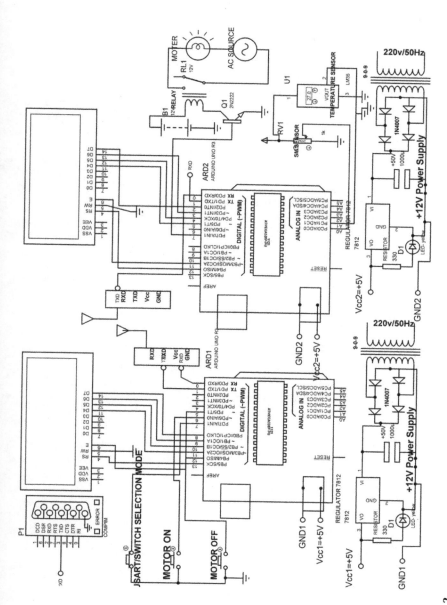

FIGURE 25.2

A circuit diagram of the system.

25.3 Program

25.3.1 Remote Control

```
#include <LiquidCrystal.h>
LiquidCrystal lcd(13, 12, 11, 10, 9, 8);
#define BUTTON_MODE_SELECTION 7
#define BUTTON_PIN_MOTORON 6
#define BUTTON_PIN_MOTOROFF 5
String SM_TEMP_STRING = "";  // a string to hold incoming data
boolean SM_TEMP_stringComplete = false;  // whether the string
is complete
void setup()
{
 lcd.begin(20,4);
 Serial.begin(9600);
 pinMode(BUTTON_MODE_SELECTION,INPUT_PULLUP);
 pinMode(BUTTON_PIN_MOTORON,INPUT_PULLUP);
 pinMode(BUTTON_PIN_MOTOROFF,INPUT_PULLUP);
 SM_TEMP_STRING.reserve(200);
 lcd.setCursor(0,0);
 lcd.print("WIRELESS IRREGATION ");
 lcd.setCursor(0,1);
 lcd.print("SYSTEM at UPES......");
 delay(1000);
}
void loop()
{
 int BUTTON_MODE_SELECTION_digital=digitalRead(BUTTON_MODE_
 SELECTION);
 if (BUTTON_MODE_SELECTION_digital==LOW)
   {
    Serial.println(30);
     if (SM_TEMP_stringComplete)
     {
       lcd.clear();
       lcd.setCursor(0,3);
       lcd.print(SM_TEMP_STRING );
       delay(10);

     SM_TEMP_STRING  = "";
     SM_TEMP_stringComplete = false;
     }
   }
```

```
else if(BUTTON_MODE_SELECTION_digital==HIGH)
        {
            int BUTTON_PIN_MOTORON_digital=digitalRead(BUTTON_
            PIN_MOTORON);
            int BUTTON_PIN_MOTOROFF_digital=digitalRead(BUTTON_
            PIN_MOTOROFF);
            if(BUTTON_PIN_MOTORON_digital==LOW)
              {
               lcd.clear();
               lcd.setCursor(0,2);
               lcd.print("MOTOR ON");
               Serial.println(10);
              }

            if(BUTTON_PIN_MOTOROFF_digital==LOW)
              {
               lcd.clear();
               lcd.setCursor(0,2);
               lcd.print("MOTOR OFF");
               Serial.println(20);
              }

        }

}

  void serialEvent()
    {
     while (Serial.available())
     {
      char inChar = (char)Serial.read();
      SM_TEMP_STRING  += inChar;
     if (inChar == 0x0D)
     {
     SM_TEMP_stringComplete = true;
     }
    }
    }
```

25.3.2 Sensor Node

```
#include <LiquidCrystal.h>
LiquidCrystal lcd(12, 11, 5, 4, 3, 2);
#define MOTOR_RELAY 7
#define SM_SENSOR A0
#define TEMP_SENSOR A1
String SWITCH_STRING = "";        choose string to HOLD data
boolean SWITCH_stringComplete = false;
```

```
void setup()
{
 Serial.begin(9600);
 lcd.begin(20,4);
 pinMode(MOTOR_RELAY, OUTPUT);
 SWITCH_STRING.reserve(200);
 lcd.setCursor(0,0);
 lcd.print("WIRELESS IRREGATION ");
 lcd.setCursor(0,1);
 lcd.print("SYSTEM at UPES......");
 delay(1000);
}

void loop()
{

 if (SWITCH_stringComplete)
     {
         lcd.clear();
         delay(10);
             if ((SWITCH_STRING[0]=='1')&&(SWITCH_
             STRING[1]=='0'))

             {
              lcd.clear();
              lcd.setCursor(0,2);
              lcd.print("MOTOR ON");
              digitalWrite(MOTOR_RELAY, HIGH);
              delay(20);
             }

             else if ((SWITCH_STRING[0]=='2')&&(SWITCH_
             STRING[1]=='0'))
             {
              lcd.clear();
              lcd.setCursor(0,2);
              lcd.print("MOTOR OFF");
              digitalWrite(MOTOR_RELAY, LOW);
              delay(20);

             }
             else if ((SWITCH_STRING[0]=='3')&&(SWITCH_
             STRING[1]=='0'))
             {
              int ANALOG_READ_SM_SENSOR_LEVEL=analogRead(SM_
              SENSOR);
              int ANALOG_READ_SM_SENSOR_PPM=ANALOG_READ_SM_
              SENSOR_LEVEL/10;
```

```
         int ANALOG_READ_TEMP_SENSOR_
         LEVEL=analogRead(TEMP_SENSOR);
         int ANALOG_READ_TEMP_SENSOR_0C=ANALOG_READ_
         TEMP_SENSOR_LEVEL/2;

         lcd.setCursor(0,3);
         lcd.print("SMSENSOR:");
         lcd.print(ANALOG_READ_SM_SENSOR_PPM);

         lcd.setCursor(0,2);
         lcd.print("TEMP");
         lcd.print(ANALOG_READ_TEMP_SENSOR_0C);

         Serial.print("SMSENSOR:");
         Serial.print(ANALOG_READ_SM_SENSOR_PPM);
         Serial.print("TEMP:");
         Serial.println(ANALOG_READ_TEMP_SENSOR_0C);
         delay(20);
       }
   if((SWITCH_STRING[0]==0x0A))
    {
         if ((SWITCH_STRING[1]=='1')&&
         (SWITCH_STRING[2]=='0'))

         {
          lcd.clear();
          lcd.setCursor(0,2);
          lcd.print("MOTOR ON");
          digitalWrite(MOTOR_RELAY, HIGH);
          delay(20);
         }

         else if ((SWITCH_STRING[1]=='2')&&(SWITCH_
         STRING[2]=='0'))
         {
          lcd.clear();
          lcd.setCursor(0,2);
          lcd.print("MOTOR OFF");
          digitalWrite(MOTOR_RELAY, LOW);
          delay(20);

         }
          else if ((SWITCH_STRING[1]=='3')&&(SWITCH_
          STRING[2]=='0'))
         {
          int ANALOG_READ_SM_SENSOR_LEVEL=analogRead(SM_
          SENSOR);
          int ANALOG_READ_SM_SENSOR_PPM=ANALOG_READ_SM_
          SENSOR_LEVEL/10;
```

```
              int ANALOG_READ_TEMP_SENSOR_
              LEVEL=analogRead(TEMP_SENSOR);
              int ANALOG_READ_TEMP_SENSOR_0C=ANALOG_READ_
              TEMP_SENSOR_LEVEL/2;

              lcd.setCursor(0,3);
              lcd.print("SMSENSOR:");
              lcd.print(ANALOG_READ_SM_SENSOR_PPM);

              lcd.setCursor(0,2);
              lcd.print("TEMP");
              lcd.print(ANALOG_READ_TEMP_SENSOR_0C);

              Serial.print("SMSENSOR:");
              Serial.print(ANALOG_READ_SM_SENSOR_PPM);
              Serial.print("TEMP:");
              Serial.println(ANALOG_READ_TEMP_SENSOR_0C);
              delay(20);
              }

            }

    SWITCH_STRING = "";
    SWITCH_stringComplete = false;
    }
 delay(10);
 }

void serialEvent()
    {
    while (Serial.available())
      {
      char inChar = (char)Serial.read();
      SWITCH_STRING += inChar;
    if (inChar == 0x0D)
      {
    SWITCH_stringComplete = true;
      }
      }
    }
```

25.4 Proteus Simulation Model

Proteus simulation model for the system is designed by connecting all the components as described in Section 25.2. Figure 25.3 shows the Proteus simulation model with the SM sensor and the temperature sensor values. In Figure 25.4, the Proteus simulation model shows the motor in "ON" condition. In Figure 25.5, the Proteus simulation model shows the motor in "OFF" condition.

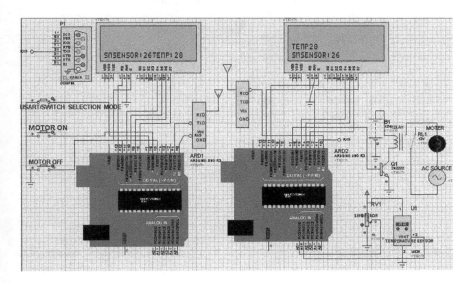

FIGURE 25.3
Proteus simulation model for the system.

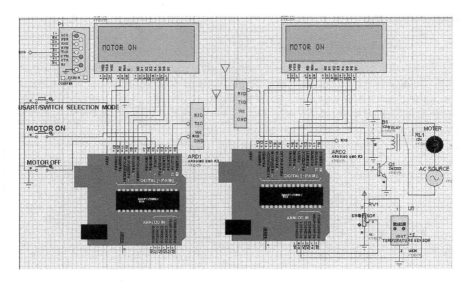

FIGURE 25.4
A Proteus simulation model showing "MOTOR ON."

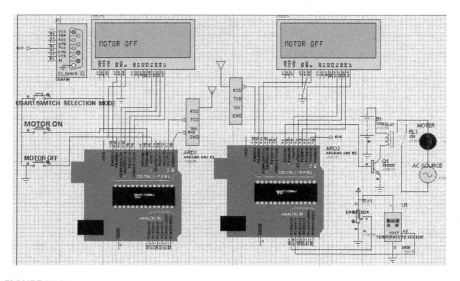

FIGURE 25.5
A Proteus simulation model showing "MOTOR OFF."

Index